# PRELIMINARY CONCEPTS

Volume 3: Equations and Inequalities in One Variable

All variables denote real numbers.

## Symbols

| | |
|---|---|
| a, b, c, A, B, C, etc. | Variables |
| R | Set of real numbers |
| ∈ | Is a member (or element) of |
| {a,b} | Set whose elements are $a$ and $b$ |
| = | Is equal to |
| ≠ | Is not equal to |
| < | Is less than |
| > | Is greater than |
| ≤ | Is less than or equal to |
| ≥ | Is greater than or equal to |
| {a \| ... } | Set of all $a$ such that ... |
| $a + b$ | Sum of $a$ and $b$ |
| $a \cdot b$, $ab$, $a(b)$, $(a)(b)$ | Product of $a$ and $b$ |
| $a - b$ | Difference of $b$ subtracted from $a$ |
| $\dfrac{a}{b}$ or $a/b$ | Quotient of $a$ divided by $b$ |
| ( ), [ ] | Parentheses, brackets; grouping devices |
| $a + b \cdot c$ | $a + b \cdot c = a + (b \cdot c)$ |
| $a \cdot b + c$ | $a \cdot b + c = (a \cdot b) + c$ |
| $a \cdot b + c \cdot d$ | $a \cdot b + c \cdot d = (a \cdot b) + (c \cdot d)$ |
| $-a$ | Additive inverse (or negative) of $a$ |
| $\dfrac{1}{a}$ $(a \neq 0)$ | Multiplicative inverse (or reciprocal) of $a$ |
| $\sqrt{a}$ $(a \geq 0)$ | Nonnegative square root of $a$ |
| $a^m$ | $a \cdot a \cdot a \cdots a$ ($m$ factors) |

## Properties

1. $a = a$ — Reflexive law of equality
2. If $a = b$, then $b = a$ — Symmetric law of equality
3. If $a = b$ and $b = c$, then $a = c$ — Transitive law of equality
4. If $a < b$ and $b < c$, then $a < c$ — Transitive law of inequality
5. If $a > 0$ and $b > 0$, then $a + b > 0$ and $a \cdot b > 0$
6. If $a = b$ then $a$ may be replaced by $b$ or $b$ may be replaced by $a$ in any statement without changing the truth or falsity of the statement. — Substitution law
7. $a + b \in R$ ⎫
8. $a \cdot b \in R$ ⎭ Closure laws
9. $a + b = b + a$ ⎫
10. $a \cdot b = b \cdot a$ ⎭ Commutative laws
11. $(a + b) + c = a + (b + c)$ ⎫
12. $(a \cdot b) \cdot c = a \cdot (b \cdot c)$ ⎭ Associative laws
13. $a + 0 = a$ ⎫
14. $a \cdot 1 = a$ ⎭ Identity laws
15. $a \cdot (b + c) = a \cdot b + a \cdot c$ ⎫
    $(b + c) \cdot a = b \cdot a + c \cdot a$ ⎭ Distributive laws
16. $a + (-a) = 0$ — Additive inverse law
17. $a \cdot \dfrac{1}{a} = 1$ $(a \neq 0)$ — Multiplicative inverse law
18. $a \cdot 0 = 0$
19. $-(-a) = a$

**Properties** (continued)

20. *To add two numbers: If the numbers have like signs, add the absolute values of the numbers and prefix to the sum the common sign of the numbers. If the numbers have unlike signs, find the nonnegative difference of the absolute values of the numbers and prefix to the difference the sign of the number with the larger absolute value.*

21. *To subtract a number $b$ from a number $a$, add the negative of $b$ to $a$:*
$a - b = a + (-b)$

22. *To multiply (or divide) two numbers, multiply (or divide) the absolute values of the numbers. If the numbers have like signs, the product (or quotient) is positive. If the numbers have unlike signs, the product (or quotient) is negative.*

23. $(+a) \cdot (+b) = a \cdot b$
24. $(-a) \cdot (-b) = a \cdot b$
25. $(+a) \cdot (-b) = (-a) \cdot (+b) = -(a \cdot b)$
26. $-a = -1 \cdot a$
27. $\dfrac{a}{b} = a \cdot \dfrac{1}{b} \quad (b \neq 0)$
28. $\dfrac{a \cdot c}{b \cdot c} = \dfrac{a}{b} \quad (b, c \neq 0)$
29. $\dfrac{a}{c} + \dfrac{b}{c} = \dfrac{a+b}{c} \quad (c \neq 0)$
30. $\dfrac{a}{b} = \dfrac{-a}{b} = -\dfrac{a}{-b} = \dfrac{-a}{-b} \quad (b \neq 0)$
31. $-\dfrac{a}{b} = \dfrac{-a}{b} = \dfrac{a}{-b} = -\dfrac{-a}{-b} \quad (b \neq 0)$
32. $\dfrac{a}{c} - \dfrac{b}{c} = \dfrac{a-b}{c} \quad (c \neq 0)$
33. $\dfrac{a}{b} \cdot \dfrac{c}{c} = \dfrac{a \cdot c}{b \cdot c} \quad (b, c \neq 0)$
34. $1 \div \dfrac{a}{b} = \dfrac{1}{\dfrac{a}{b}} = \dfrac{b}{a} \quad (a, b \neq 0)$
35. $\dfrac{a}{b} \div \dfrac{c}{d} = \dfrac{a}{b} \cdot \dfrac{d}{c} \quad (b, c, d \neq 0)$
36. $\sqrt{a} \cdot \sqrt{a} = \sqrt{a^2} = a \quad (a \geq 0)$
37. $\sqrt{ab} = \sqrt{a}\sqrt{b} \quad (a, b \geq 0)$
38. $x^2 + (a+b)x + ab = (x+a)(x+b)$
39. $x^2 + 2ax + a^2 = (x+a)^2$
40. $x^2 - a^2 = (x+a)(x-a)$

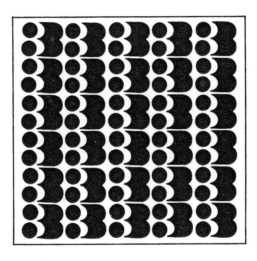

# EQUATIONS AND INEQUALITIES IN ONE VARIABLE

# CHARLES C. CARICO

Los Angeles Pierce College

Consulting Editor

Irving Drooyan

Los Angeles Pierce College

Wadsworth Publishing Company, Inc.
Belmont, California

*Designers: Wadsworth Design Staff*

*Mathematics Editor: Don Dellen*

*Production Editor: Dorothy Graham*

*Copy Editor: Alice Goehring*

*Technical Illustrator: Carleton Brown*

© 1974 by Wadsworth Publishing Company, Inc., Belmont, California 94002. All rights reserved. No part of this book may be reproduced, stored in a retrieval system or transcribed, in any form or by any means, electronic, mechanical, photocopying, recording or otherwise, without the prior written permission of the publisher.

*ISBN 0-534-00313-3*
*L. C. Cat. Card No. 73-86310*
*Printed in the United States of America*

*1 2 3 4 5 6 7 8 9 10—78 77 76 75 74*

# PREFACE

*About the series*

This is the third volume in the Wadsworth precalculus mathematics series for college students. The titles of the twelve volumes of the series are listed on the back cover. Selected volumes in the series can be used (1) as the basic textbook for any standard precalculus course, (2) for independent study as a supplement to another text used in the course, or (3) for a self-study review of material previously studied by a student.

The modular organization and following features of this series make it particularly useful for independent work or to accommodate individual differences in formal courses.

1. Beginning with the second volume, each volume includes lists of all the mathematical symbols and properties covered in previous volumes that are necessary prerequisites for that particular volume.
2. The textual material in each section is presented in a concise style to highlight the basic mathematical ideas and instructional objectives of the section.
3. Each group of exercises covering a particular topic is preceded by detailed examples of a similar type.
4. A detailed Solution Key for all exercises is included in each volumes 1 to 8. Solution Keys for the odd-numbered exercises are included in volumes 9 to 12.

In formal lecture presentations, it is anticipated that instructors will indicate how the material in a particular section relates to the preceding and subsequent topics in the volume, and that they will introduce mathematical structure, from elementary heuristic motivations to an axiomatic approach, commensurate with the objectives of the course. Instructors may wish to make a distinction between axioms and theorems, which we have listed together as properties. Also, they may wish to prove some of the theorems. Because of the detailed examples and complete Solution Key contained in each volume, lectures for each section can be brief. Hence, if facilities are available, these lectures are readily adaptable to short videotape presentations.

The list of the volumes in this series presented on the back cover shows the number of sections and the number of review exercise sets in each volume. Since each section and the review exercise set at the end of each unit has been designed as an assignment for one class period, this information can be helpful in selecting the number of volumes that can be completed for courses of different lengths. The guide manual for this series, which is available from the publisher, includes material that can also be used to select appropriate volumes for a particular course or for self-study. This manual contains a table of contents for all volumes, all unit reviews, and a complete summary of all the symbols and properties introduced in the series.

*About this volume*

This volume is concerned with the mathematics necessary to solve first- and second-degree equations and inequalities in one variable. Also, attention is given to the important tasks of translating word statements and sentences into symbolic form and solving word problems. The instructional objectives of each unit can be readily observed from three sources: the table of contents; the list of symbols and properties on the back endpaper, which summarizes the topics introduced in these units; and the review exercises at the end of each unit.

The following preface to students includes our suggestions as to how students can use this text.

CHARLES C. CARICO

# TO THE STUDENT

The organization of the material in this volume probably differs from that of other mathematics textbooks you may have used. Therefore, we shall make some suggestions as to how to use this volume, whether you are using it as the basic textbook for your course, for independent study as a supplement to another text used in your class, or for review of material that you have studied previously.

To begin, acquaint yourself with the overall organization of this volume:

1. Mathematical symbols and properties that are prerequisites for the topics introduced in this volume are listed on the front endpaper. If many of these symbols and properties appear totally unfamiliar to you, you should probably review previous volumes in this series before beginning this volume.
2. A Solution Key, containing a complete solution for each exercise, is included at the end of the volume.
3. Each section begins with a list of the definitions and notation and a list of the properties that are introduced and used in that section. The properties include both the assumptions that we make about numbers and operations and some logical consequences of the assumptions.
4. Each unit concludes with a set of review exercises which are representative of the exercises throughout the unit. This review may conveniently be used as a self-test for the unit or as a diagnostic test to determine if you need to study a particular unit.
5. A summary of the symbols and properties introduced in this volume is included on the back endpaper.

After you have become familiar with the organization of the material, we suggest that before starting a unit, you determine the instructional objectives you expect to have attained upon completing it. These objectives can be determined from the references noted in 3, 4, and 5 above.

When beginning each section, it is suggested that you give the introductory material a preliminary reading. It is not expected that the definitions, symbols, and properties will be very meaningful at this first reading. As you proceed to work the exercises, you will probably want to refer to this material frequently. After you complete the assignment, you should reread the introductory material, this time with greater understanding. The detailed examples in each set of exercises should be studied carefully before proceeding to the exercises that follow. The format presented in these examples should be closely followed in your work. As you work the exercises you may wish to refer to the symbols and properties listed on the front endpaper (which we assume you have encountered in a previous course). However, no doubt you have forgotten some of these, and so probably you will need to refer to this list frequently as you begin this volume but less frequently as you proceed through it.

The Solution Key can be used in various ways. You may want to check your answer to each exercise as it is completed, or perhaps you would rather check the answers after you have completed an entire exercise set. Remember that many exercises can be solved in different ways. In order to keep the Solution Key to a convenient size, alternative solutions are not shown. We anticipate that sometimes you will devise ways to solve problems that are simpler than ours. However, if you have difficulty with a particular exercise, you should refer to the solution given in the Solution Key to see one way in which the exercise can be solved. (Of course, it is best to do this only after you have made an honest attempt to work the problem yourself.)

As previously suggested, upon completing a unit, you might want to evaluate your progress by considering the Unit Review as a self-test. If you do so, you should complete the entire review before checking your answers. The results (you may wish to give yourself a grade) should suggest whether you are ready to proceed to the next unit or whether you should review a particular section or sections before proceeding.

When you have completed this volume, you may want to detach the summary page of the symbols and properties on the back endpaper as a convenient reference chart for your further studies.

This volume has been prepared in such a way as to make *you* responsible for your success. Good luck!

CHARLES C. CARICO

# CONTENTS

**Unit 1 First-degree Equations and Inequalities 1**

    1.1 Solution of Equations: I    1
    1.2 Solution of Equations: II    5
    1.3 Solving Equations for Specified Symbols    8
    1.4 Inequalities    10
    1.5 Word Problems    13
         Unit Review    21

**Unit 2 Second-degree Equations 22**

    2.1 Solution by Factoring    22
    2.2 Solutions of Equations of the Form $x^2 = c$    25
    2.3 Solution by Completing the Square    26
    2.4 Solution by the Quadratic Formula    28
    2.5 Word Problems    30
         Unit Review    33

**Appendix: Quadratic Inequalities 34**

**Solution Key 37**

**Index 80**

# UNIT 1

# FIRST-DEGREE EQUATIONS AND INEQUALITIES

Many practical problems are concerned with numerical measures of physical quantities. In this volume we will consider methods of solving such problems when their solutions are not immediately evident. We begin by solving simple first-degree equations and first-degree inequalities, and then we will use them as mathematical models to solve word problems.

## 1.1 SOLUTION OF EQUATIONS: I

### Definitions and Notation

**Equation**     A statement that two quantities are equal; the statement may be true or false, or it may not be possible to tell

*Examples*

a. $2 + 3 = 5$ is an equation that is a true statement.
b. $2 + 8 = 9$ is an equation that is a false statement.
c. $x + 4 = 7$ is an equation whose truth or falsity depends upon the number replacement for the variable.

**Solution or root**     A replacement for the variable in an equation resulting in a true statement

**Solution set**     The set of all solutions of an equation

*Examples*

a. The solution or root of $x + 3 = 7$ is 4 because $(4) + 3 = 7$ is a true statement. Since 4 is the only solution, the solution set is {4}.
b. The solution set of $5 = x - 8$ is {13}, because $5 = (13) - 8$ is a true statement and 13 is the only solution of the equation.

| | |
|---|---|
| **To solve an equation** | To find the solution set of the equation |
| **Unknown** | A name for the variable in an equation |
| **Identity** | An equation that is true for all possible replacements of the variables |
| **Conditional equation** | An equation that is a true statement for at least one replacement, but not all replacements, of the variable for which each member is defined |
| **Formula** | An equation derived from a practical problem |

*Examples*

a. $x \cdot 1 = x$ is an *identity* because any real number replacement for $x$ results in a true statement.
b. $x + 3 = 7$ is a *conditional equation* because a true statement is not obtained for all real number replacements for $x$, but a true statement is obtained for at least one real number replacement for $x$.
c. $A = \pi r^2$ and $C = 2\pi r$ are called formulas because they are equations which are derived from practical problems.

| | |
|---|---|
| **Equivalent equations** | Equations that have identical solution sets |

*Examples*

a. $x + 4 = 7$ and $x = 3$ are equivalent equations because they have identical solution sets, {3}.
b. $4x = 16$, $3x = 12$, $2x = 8$, and $x = 4$ are equivalent equations because they have identical solution sets, {4}.

| | |
|---|---|
| **Member of an equation** | Each of the two expressions on either side of the equal sign in the equation |
| **$P(x), Q(x), R(x)$, etc.** | Names of polynomials; these symbols can represent "constant polynomials," that is, any real numbers such as 2, $-8$, and 2/3 |

## 1.1 Solution of Equations: I

*Examples*

a. In the equation $x + 4 = 7$, $x + 4$ is the left-hand member and 7 is the right-hand member.
b. If $P(x) = 2x + 1$ and $Q(x) = x^2 + 5$, then
$$P(x) + Q(x) = (2x + 1) + (x^2 + 5).$$

## Properties*

The equations $P(x) = Q(x)$ and each one of

| | |
|---|---|
| $P(x) + R(x) = Q(x) + R(x)$ | Addition law of equality |
| $P(x) - R(x) = Q(x) - R(x)$ | Subtraction law of equality |
| $P(x) \cdot R(x) = Q(x) \cdot R(x) \quad (R(x) \neq 0)$ | Multiplication law of equality |
| $\dfrac{P(x)}{R(x)} = \dfrac{Q(x)}{R(x)} \quad (R(x) \neq 0)$ | Division law of equality |

are equivalent.

An equivalent equation is produced by

1. Adding the same polynomial to, or subtracting the same polynomial from, each member of an equation, or
2. Multiplying or dividing each member of an equation by the same polynomial representing a nonzero number.

## Exercises

*Determine if the given number is a solution of the given equation*

Example     $3y + 6 = 4y - 4;\quad 10$

Solution     Substitute 10 for $y$ and simplify each member
Does   $3(10) + 6 = 4(10) - 4$?
Does   $30 + 6 = 40 - 4$?
Does   $36 = 36$?
Yes.

1. $x - 3 = 7;\quad 4$
2. $2x - 6 = 3;\quad 4$
3. $2x + 6 = 3x - 5;\quad 3$
4. $5x - 1 = 2x + 2;\quad 1$
5. $3x + 2 = 8 + x;\quad 3$
6. $3x - 5 = 2x + 7;\quad -1$
7. $3 = 6x + 3(x - 2);\quad -1$
8. $6 - 2x = 6(2x + 1);\quad 0$

*Throughout this volume it is assumed that all variables denote real numbers unless otherwise noted.

**4**   Unit 1   First-degree Equations and Inequalities

9. $0 = 6x - 24$;   4
10. $0 = 7x + 7$;   $-1$
11. $\frac{1}{4}x - 3 = x + 2$;   8
12. $\frac{1}{3}x - 2x = -12 + \frac{1}{3}x$;   6

*Solve. If the solution set is obvious, write it directly. If not, generate equivalent equations until the solution set is obvious.*

Example   $\frac{x}{4} = 3$

Solution   If the solution set {12} is not obvious, multiply each member by 4.

$$4\left(\frac{x}{4}\right) = 4(3)$$

$$x = 12$$

The solution set {12} should now be evident by inspection.

13. $x + 2 = 5$
14. $x - 3 = 7$
15. $12 = x + 7$
16. $6 = x - 3$
17. $2x = 8$
18. $7x = 21$
19. $-12 = -3x$
20. $-6 = 2x$
21. $-\frac{x}{3} = 5$
22. $\frac{x}{6} = 2$
23. $\frac{x}{4} = -1$
24. $\frac{x}{-6} = -3$

Example   $4x + 3 = 2x - 7$

Solution   By adding $-2x - 3$ to each member, we obtain an equation that has only terms containing $x$ in one member and only constant terms in the other.

$$4x + 3 + (-2x - 3) = 2x - 7 + (-2x - 3)$$

Notice that after adding the $-2x$, the $x$ terms in the right-hand member add to 0. Similarly, after adding $-3$, the constant terms in the left-hand member add to 0.

$$2x = -10$$

Divide each member by 2, the coefficient of $x$.

$$\frac{2x}{2} = \frac{-10}{2}$$

$$x = -5$$

The solution set is $\{-5\}$.

25. $x = 3 - 2x$
26. $2x = 3 + x$
27. $4 - x = x + 2$
28. $2 - x = 4 + 3x$
29. $3 = x - 4 + 6x$
30. $3x + 2 = 4x - 6 + x$

*Solve each formula for the specified variable.*

31. $E = IR$;   $I$, given that   $E = 2.22$   and   $R = 0.60$.
32. $C = \pi D$;   $D$, given that   $C = 42.31$   and   $\pi = 3.14$.
33. $P = 2l + 2w$;   $w$, given that   $P = 42$   and   $l = 6$.
34. $A = P(l + r)$;   $P$, given that   $A = 642$   and   $r = 0.04$.
35. $A = \pi r^2 h$;   $h$, given that   $A = 46.12$   and   $r = 2.00$   (use $\pi = 3.14$).
36. $A = \pi r^2 + 2\pi rh$;   $h$, given that   $A = 235.5$   and   $r = 3.0$   (use $\pi = 3.14$).

## 1.2 SOLUTION OF EQUATIONS: II

### Exercises

*Solve each equation.*

**Example**   $6(x - 2) - (x + 1) = 2$

**Solution**   Apply the Distributive law.

$$6x - 12 - x - 1 = 2$$

Combine like terms.

$$5x - 13 = 2$$

Add 13 to each member.

$$5x - 13 + 13 = 2 + 13$$
$$5x = 15$$

Divide each member by 5, the coefficient of $x$.

$$\frac{5x}{5} = \frac{15}{5}$$

$$x = 3$$

The solution set is {3}.

1. $2(x + 5) = 16$
2. $3(x + 1) = 3$
3. $6 = 2(2x - 1)$
4. $3 = 4(2x + 1)$
5. $x - (8 - x) = 2$
6. $2x - (3 - x) = 0$
7. $3(7 - 2x) = 30 - 7(x + 1)$
8. $6x - 2(2x + 5) = 6(5 + x)$
9. $-3[x - (2x + 3) - 2x] = -9$
10. $5[2 + 3(x - 2)] + 20 = 0$
11. $-2[x - (x - 1)] = -3(x + 1)$
12. $3[2x - (x + 2)] = -3(3 - 2x)$

## Unit 1 First-degree Equations and Inequalities

**Example**  $4 - (x - 1)(x + 2) = 8 - x^2$

**Solution**  Apply the Distributive law.
$$4 - (x^2 + x - 2) = 8 - x^2$$
$$4 - x^2 - x + 2 = 8 - x^2$$

Add $x^2 - 6$ to each member.
$$6 - x^2 - x + (x^2 - 6) = 8 - x^2 + (x^2 - 6)$$
$$-x = 2$$

Multiply or divide each member by $-1$, the coefficient of $x$.
$$x = -2$$

The solution set is $\{-2\}$.

13. $(x - 1)(x - 1) = x^2 - 11$
14. $(2x + 1)(x - 3) = (x - 2)(2x + 1)$
15. $(x - 2)^2 = x^2 - 8$
16. $(x - 1)^2 = x^2 - 15$
17. $4 - (x - 3)(x + 2) = 10 - x^2$
18. $0 = x^2 + (2 - x)(x + 5)$

*Alternate methods of solutions are possible for many of the exercises that follow. Generally it is easier to solve an equation that contains fractions by first generating an equivalent equation without fractions by multiplying each member by the least common denominator of all the fractions in the original equation.*

**Example**  $\dfrac{x}{3} + 4 = x - 2$

**Solution**  We first eliminate the fraction. Since the denominator is 3, multiply *each* member by 3 and then simplify.
$$3\left(\frac{x}{3} + 4\right) = 3(x - 2)$$
$$3\left(\frac{x}{3}\right) + 12 = 3x - 6$$
$$x + 12 = 3x - 6$$

Add $-x + 6$ to each member.
$$x + 12 + (-x + 6) = 3x - 6 + (-x + 6)$$
$$18 = 2x$$

Divide each member by 2, the coefficient of $x$.
$$\frac{18}{2} = \frac{2x}{2}$$
$$9 = x$$

The solution set is $\{9\}$.

## 1.2 Solution of Equations: II

**19.** $\dfrac{x}{3} = 2$  **20.** $\dfrac{2x}{3} = -4$  **21.** $\dfrac{5x - x}{6} = 4$

**22.** $7 - \dfrac{x}{3} = x - 1$  **23.** $6 - \dfrac{x}{2} = x$  **24.** $\dfrac{2x}{3} - 3 = x - 5$

**Example**  $\dfrac{2x}{3} - x = \dfrac{5}{2}$

**Solution**  We first eliminate the fractions. Determine the least common denominator of $\dfrac{2x}{3}$ and $\dfrac{5}{2}$ ($x$ can be viewed as $\dfrac{x}{1}$) and then multiply both members of the equation by the least common denominator. The least common denominator is 6.

$$6\left(\dfrac{2x}{3} - \dfrac{x}{1}\right) = 6\left(\dfrac{5}{2}\right)$$

$$4x - 6x = 15$$

Combine like terms.

$$-2x = 15$$

Divide each member by $-2$, the coefficient of $x$.

$$\dfrac{-2x}{-2} = \dfrac{15}{-2}$$

$$x = -\dfrac{15}{2}$$

The solution set is $\left\{-\dfrac{15}{2}\right\}$.*

**25.** $2 + \dfrac{x}{6} = \dfrac{5}{2}$  **26.** $1 + \dfrac{x}{9} = \dfrac{4}{3}$  **27.** $\dfrac{x}{5} - \dfrac{x}{2} = 9$

**28.** $\dfrac{y}{2} + \dfrac{y}{3} - \dfrac{y}{4} = 7$  **29.** $\dfrac{2x - 1}{5} = \dfrac{x + 1}{2}$  **30.** $\dfrac{2x}{3} - \dfrac{2x + 5}{6} = \dfrac{1}{2}$

**Example**  $\dfrac{2}{3} = 6 - \dfrac{x + 10}{x - 3}$

**Solution**  Determine the least common denominator, $3(x - 3)$. Multiply each member of the equation by $3(x - 3)$.

$$3(x - 3)\dfrac{2}{3} = 3(x - 3)6 - 3(x - 3)\dfrac{x + 10}{x - 3}$$

$$2(x - 3) = 18(x - 3) - 3(x + 10) \qquad \text{(continued)}$$

---

* We will use either of the forms $-\dfrac{a}{b}$ or $\dfrac{-a}{b}$ to indicate a negative fraction.

Rewrite without parentheses.
$$2x - 6 = 18x - 54 - 3x - 30$$
Combine like terms and add $-15x + 6$ to each member.
$$2x - 6 + (-15x + 6) = 15x - 84 + (-15x + 6)$$
$$-13x = -78$$
Divide each member by $-13$.
$$x = 6$$
The solution set is $\{6\}$.

31. $\dfrac{3}{5} = \dfrac{x}{x+2}$

32. $\dfrac{2}{x-2} = \dfrac{9}{x+12}$

33. $\dfrac{x}{x-2} = \dfrac{2}{x-2} + 2$

34. $\dfrac{5}{x-3} = \dfrac{x+2}{x-3} + 2$

35. $\dfrac{2}{y+1} + \dfrac{1}{3y+3} = \dfrac{1}{6}$

36. $\dfrac{y}{y+2} - \dfrac{3}{y-2} = \dfrac{y^2+8}{y^2-4}$

*Solve each formula for the specified variable.*

37. $F = \dfrac{9}{5}C + 32$;  $C$, given that  $F = 176$.

38. $s - \dfrac{1}{2}gt^2 = 0$;  $s$, given that  $g = 32.2$ and $t = 4$.

39. $S = \dfrac{a}{1-r}$;  $r$, given that  $a = 6$ and  $S = 9$.

40. $A = \dfrac{h}{2}(b + c)$;  $b$, given that  $A = 240$, $h = 20$, and $c = 12$.

## 1.3 SOLVING EQUATIONS FOR SPECIFIED SYMBOLS

The methods used in the preceding section to generate equivalent equations to find numerical solutions to equations in one variable can also be used to generate equivalent equations if the equations contain more than one variable.

## Exercises

*Solve for x or y. (Leave the results in the form of an equation.) Assume that no denominator equals 0.*

Example        $2ax = b - 3ax$

## 1.3 Solving Equations for Specified Symbols

Solution: Add $3ax$ to each member so that the left-hand member will contain only terms having $x$ as a factor and the right-hand member will contain all other terms.

$$2ax + (3ax) = b - 3ax + (3ax)$$
$$5ax = b$$

Divide each member by $5a$, the coefficient of $x$.

$$\frac{5ax}{5a} = \frac{b}{5a}$$

$$x = \frac{b}{5a}$$

1. $ax = b + c$
2. $b - c = ay$
3. $3ay - b = c$
4. $4by + c = b$
5. $ax = b - 2ax$
6. $4by = a + 2by$
7. $\frac{a}{b}y + c = 0$
8. $b = \frac{1}{c}y + a$
9. $\frac{x+a}{b} = c$
10. $b = \frac{y-a}{c}$
11. $\frac{c-2y}{b} = a$
12. $\frac{b-3x}{a} = c$

Example: $5by - 2a = 2ay$

Solution: Add $2a - 2ay$ to each member so that the left-hand member will contain only terms having $y$ as a factor and the right-hand member will contain all other terms.

$$5by - 2a + (2a - 2ay) = 2ay + (2a - 2ay)$$
$$5by - 2ay = 2a$$

Factor the left member.

$$y(5b - 2a) = 2a$$

Divide each member by $5b - 2a$, the coefficient of $y$.

$$\frac{y(5b - 2a)}{5b - 2a} = \frac{2a}{5b - 2a}$$

$$y = \frac{2a}{5b - 2a}$$

13. $ax = a - x$
14. $y = b + by$
15. $a(a - x) = b(b - x)$
16. $4y - 3(y - b) = 8b$
17. $(x - 4)(b + 2) = b$
18. $(y - 2)(a + 3) = 2a$
19. $\frac{2}{x} + \frac{2}{a} = 6$
20. $\frac{2}{y} - \frac{3}{b} = 4$
21. $\frac{a}{y} - \frac{1}{2} = \frac{a}{2y}$
22. $\frac{c}{3} + \frac{1}{x} = \frac{c}{6x}$
23. $\frac{1}{a} + \frac{1}{b} = \frac{1}{x}$
24. $\frac{1}{a} - \frac{1}{b} = \frac{1}{x}$

Example     $v = k + gt$, for $g$

Solution    Add $-k$ to (or subtract $k$ from) each member so that the right-hand member will contain only terms having $g$ as a factor and the left-hand member will contain all other terms.

$$v + (-k) = k + gt + (-k)$$
$$v - k = gt$$

Divide each member by $t$, the coefficient of $g$.

$$\frac{v-k}{t} = \frac{gt}{t}$$

$$\frac{v-k}{t} = g \quad \text{or} \quad g = \frac{v-k}{t}$$

25. $v = k + gt$, for $k$    26. $E = mc^2$, for $m$    27. $f = ma$, for $m$
28. $I = prt$, for $p$    29. $pv = K$, for $v$    30. $E = IR$, for $R$
31. $s = \frac{1}{2}at^2$, for $a$    32. $p = 2l + 2w$, for $l$    33. $v = k + gt$, for $t$
34. $y = \frac{k}{z}$, for $z$    35. $V = lwh$, for $h$    36. $W = I^2R$, for $R$
37. $180 = A + B + C$, for $B$    38. $S = \frac{a}{1-r}$, for $r$
39. $A = \frac{h}{2}(b + c)$, for $c$    40. $S = 2r(r + h)$, for $h$
41. $S = 3\pi d + 5\pi D$, for $d$    42. $A = 2\pi rh + 2\pi r^2$, for $h$

# 1.4 INEQUALITIES

## Definitions

| | |
|---|---|
| **Inequality** | A statement that one quantity is less than, or greater than, another |
| **Solution of an inequality** | A replacement for a variable in an inequality resulting in a true statement |
| **Solution set of an inequality** | The set of all solutions of an inequality |

## 1.4 Inequalities

*Examples*

a. One solution of the inequality $x > 4$ is 5 because $5 > 4$ is a true statement.
b. The solution set of $x > 4$ is $\{x \mid x > 4\}$ because a true statement results if $x$ is replaced by any member of the set.

**To solve an inequality**  To find the solution set

**Equivalent inequalities**  Inequalities that have identical solution sets

*Example*

$x + 3 > 7$ and $x > 4$ are equivalent inequalities because they have identical solution sets, $\{x \mid x > 4\}$.

# Properties

The inequalities $P(x) < Q(x)$ and each one of

$P(x) + R(x) < Q(x) + R(x)$    Addition law of inequality
$P(x) - R(x) < Q(x) - R(x)$    Subtraction law of inequality

$\left. \begin{array}{l} P(x) \cdot R(x) < Q(x) \cdot R(x) \text{ if } R(x) > 0 \\ P(x) \cdot R(x) > Q(x) \cdot R(x) \text{ if } R(x) < 0 \end{array} \right\}$ Multiplication laws of inequality

$\left. \begin{array}{l} \dfrac{P(x)}{R(x)} < \dfrac{Q(x)}{R(x)} \text{ if } R(x) > 0 \\ \dfrac{P(x)}{R(x)} > \dfrac{Q(x)}{R(x)} \text{ if } R(x) < 0 \end{array} \right\}$ Division laws of inequality

are equivalent.

*An equivalent inequality is produced by*

1. *Adding the same polynomial to, or subtracting the same polynomial from, each member of an inequality, or*
2. *Multiplying or dividing each member of an inequality by the same positive number, or*
3. *Multiplying or dividing each member of an inequality by the same negative number and reversing the sense, or order, of the inequality ($a < b$ changed to $a > b$, or $a > b$ changed to $a < b$).*

# Exercises

*Solve each inequality and represent the solution set on a line graph.*

Examples   a. $\dfrac{x}{2} - 1 < 2$   b. $\dfrac{x + 4}{3} \leq 6 + x$

**Solutions**   **a.** Add 1 to each member.

$$\frac{x}{2} < 3$$

Multiply each member by 2.

$$x < 6$$

The solution set is $\{x \mid x < 6\}$.

(The open dot indicates that 6 *is not* a solution.)

**b.** Multiply each member by 3.

$$x + 4 \leq 18 + 3x$$

Add $-4 - 3x$ to each member.

$$-2x \leq 14$$

Divide each member by $-2$ and *reverse* the sense of the inequality.

$$x \geq -7$$

The solution set is $\{x \mid x \geq -7\}$.

(The solid dot indicates that $-7$ *is* a solution.)

1. $3x < 6$
2. $2x > 8$
3. $x + 7 \geq 8$
4. $x - 5 \leq 7$
5. $-3x > 12$
6. $-4x < 20$
7. $2x - 3 < 4$
8. $3x + 2 > 8$
9. $3 - 2x > 12$
10. $4 - 3x < 13$
11. $2 - x < -3$
12. $3 - x > -4$
13. $3x - 2 \geq 1 + 2x$
14. $2x + 3 \leq x - 1$
15. $\dfrac{2x - 6}{3} > 0$
16. $\dfrac{2x - 3}{2} \leq 0$
17. $\dfrac{5x - 7x}{3} \geq 0$
18. $\dfrac{x - 3x}{5} \leq 6$
19. $\dfrac{R + 2}{3} < \dfrac{R - 2}{4}$
20. $\dfrac{2R}{5} > R - \dfrac{1}{2}$
21. $\dfrac{T - 1}{3} \geq 2 + \dfrac{T}{2}$
22. $\dfrac{2 - T}{5} \leq -\dfrac{8 + T}{3}$
23. $\dfrac{P - 2P}{4} < \dfrac{2P}{3} + \dfrac{1}{2}$
24. $\dfrac{P}{2} - \dfrac{2P}{3} > \dfrac{P}{4}$

## 1.5 WORD PROBLEMS

## Suggestions for Solving Word Problems

1. *Represent the quantities asked for by symbols. In this section use only one variable for all relevant quantities.*
2. *Where applicable, draw a figure and label all known quantities and all unknown quantities using symbols.*
3. *Write an equation which expresses the condition on the variable.*
4. *Solve the equation and satisfy yourself that each solution meets the requirements of the original problem.*

## Exercises

(a) Set up an equation for each problem
(b) Solve.

*General Problems*

Example    The first stage of a rocket burns 28 seconds longer than the second stage. If the total burning time for both stages is 152 seconds, how long does each stage burn?

Solution    (a) Represent the unknown quantities using symbols.
Let $t$ = burning time of the first stage; then
$t - 28$ = burning time of second stage.
Write an equation relating the unknown quantities.

$$t + (t - 28) = 152$$

(b) Solve the equation.

$$t + t - 28 = 152$$
$$2t - 28 = 152$$
$$2t = 180$$
$$t = 90;$$
$$t - 28 = 62$$

Therefore, the first stage burns 90 seconds and the second stage 62 seconds.

1. One number is 2 less than three times another. Find the numbers if their sum is 30.
2. If $\frac{3}{5}$ of a number is 8 less than the number, what is the number?
3. A 40-foot rope is cut into two pieces, one of which is 2 feet shorter than six times the length of the other. How long is each piece of rope?
4. In an election, 7179 votes were cast for two candidates. If 6 votes had switched from the winner to the loser, the loser would have won by a single vote. How many votes were cast for each candidate?
5. One side of a long-playing record contains 5 songs, with 15-second intervals between songs. If all of the songs take an equal amount of time to play, and if the entire side takes 26 minutes to play, how many minutes does each song last?
6. In a 30-minute TV program, the entertainment portion lasts 2 minutes longer than three times the duration of the commercial portion. How much time is devoted to commercials during the 30 minutes?

*Consecutive-integer Problems*

Solving consecutive-integer problems ordinarily involves using the fact that consecutive integers differ by 1; that is, consecutive integers can be represented by $x, x+1, x+2, \ldots$, etc. Of course, consecutive even integers, or consecutive odd integers, are represented by $x, x+2, x+4, \ldots$, etc.

Example   The sum of three consecutive integers is 21 greater than two times the smallest of the three integers. Find the integers.

Solution   (a) Represent the integers using symbols.
Let $x =$ an integer; then
$x + 1 =$ next consecutive integer,
$x + 2 =$ next consecutive integer.

Write an equation representing the conditions given in the problem.
$$x + (x + 1) + (x + 2) = 2x + 21$$
(b) Solve the equation.
$$3x + 3 = 2x + 21$$
$$x = 18;$$
$$x + 1 = 19$$
$$x + 2 = 20$$
Therefore, the integers are 18, 19, and 20.

7. Find three consecutive integers whose sum is 42.
8. The sum of two consecutive integers is 8 less than three times the lesser integer. Find the integers.

## 1.5 Word Problems

9. If ⅓ of an integer is added to ½ the next consecutive integer, the sum is 33. Find the integers.
10. If ½ of an integer is added to ⅓ the next consecutive integer, the sum is 17. Find the integers.
11. The sum of three consecutive odd integers is 63. Find the integers.
12. If the lesser of two consecutive even integers is multiplied by $-4$, the product is 46 more than the sum of the integers. Find the integers.

*Age Problems*

Problems involving ages are similar to integer problems, because, ordinarily, ages are considered in whole numbers of years. In solving age problems, it is often necessary to express a person's (or thing's) age at various times in terms of his (or its) age now. The table below illustrates a few such cases.

| If age today is | then age 5 years ago was | and age 10 years from today will be |
|---|---|---|
| $x$ | $x - 5$ | $x + 10$ |
| $y - 3$ | $y - 8$ | $y + 7$ |
| $z + 2$ | $z - 3$ | $z + 12$ |

**Example**  A man is presently nine times as old as his son. In 9 years, he will be only three times as old as his son. How old is each now?

**Solution**  
(a) Represent the unknown quantities symbolically.
Let $x =$ son's age now; then
$9x =$ father's age now.
In 9 years, $x + 9 =$ son's age, and $9x + 9 =$ father's age.
Write an equation relating the ages 9 years hence.
$$9x + 9 = 3(x + 9)$$
(b) Solve the equation.
$$9x + 9 = 3x + 27$$
$$6x = 18$$
$$x = 3;$$
$$9x = 27$$

Therefore, today, the son's age is 3 years and the father's age is 27 years.

13. A boy is three times as old as his sister. In 4 years, he will be twice as old as his sister. How old is each now?
14. A woman is five times as old as her daughter. In 9 years, the woman will be three times as old as her daughter. How old is each now?

15. A boy is 4 years older than his sister. Ten years ago, he was twice as old as his sister. How old is each now?
16. A father is 25 years older than his son. In 10 years, the father will be twice as old as his son. How old is each now?
17. The sum of the ages of a father and his son is 60 years. Four years from now, the father will be three times as old as his son. How old is each now?
18. The sum of the ages of Joe and Pete is 45 years. Five years from now, Joe will be four times as old as Pete. How old is each now?

*Coin Problems*

The basic idea of problems involving coins (or bills) is that the value of a number of coins, or bills, of the same denomination is equal to the product of the value of a single coin (or bill) and the total number of coins (or bills). That is,

$$\begin{bmatrix} \text{Value of} \\ n \\ \text{coins} \end{bmatrix} = \begin{bmatrix} \text{Value of} \\ \text{one} \\ \text{coin} \end{bmatrix} \times \begin{bmatrix} \text{Number} \\ \text{of} \\ \text{coins} \end{bmatrix}$$

It is helpful in most coin problems to think in terms of cents rather than dollars. Tables can also be helpful in solving coin problems.

**Example**  A collection of coins consisting of dimes and quarters has a value of $11.60. How many dimes and quarters are in the collection if there are 32 more dimes than quarters?

**Solution**  (a) Represent the unknown quantities symbolically.
Let $x$ = number of quarters; then
$x + 32$ = number of dimes.
Make a table.

| Denomination | Value of one coin | Number of coins | Value of coins |
|---|---|---|---|
| Dimes | 10 | $x + 32$ | $10(x + 32)$ |
| Quarters | 25 | $x$ | $25x$ |

Write an equation relating the value of the quarters and value of the dimes to the value of the entire collection.

$$\begin{bmatrix} \text{Value of} \\ \text{quarters} \\ \text{in cents} \end{bmatrix} + \begin{bmatrix} \text{Value of} \\ \text{dimes} \\ \text{in cents} \end{bmatrix} = \begin{bmatrix} \text{Value of} \\ \text{collection} \\ \text{in cents} \end{bmatrix}$$

$$25x + 10(x + 32) = 1160$$

## 1.5 Word Problems

(b) Solve for $x$.

$$25x + 10x + 320 = 1160$$
$$35x = 840$$
$$x = 24;$$
$$x + 32 = 56$$

Therefore there are 24 quarters and 56 dimes in the collection.

19. A man has $1.80 in dimes and nickels with three more dimes than nickels. How many of each coin does he have?
20. A man has $2.80 in dimes and quarters. If he has seven more dimes than quarters, how many of each kind of coin does he have?
21. A collection of 16 coins consisting entirely of nickels and quarters has a value of $2.20. How many of each coin are in the collection?
22. A man has $446 in ten-, five-, and one-dollar bills. There are 94 bills in all and 10 more five-dollar bills than ten-dollar bills. How many of each kind does he have?
23. A vendor bought a supply of ice cream bars at three for 20 cents. He ate one and sold the remainder at 10 cents each. If he made $2.00 profit, how many bars did he buy?
24. The admission at a baseball game was $1.50 for adults and 85 cents for children. The receipts were $93.10 for 82 paid admissions. How many adults and children attended the game?

*Interest Problems*

Simple interest problems involve the fact that the amount of interest earned during a single year is equal to the amount invested times the rate of interest. Thus, $I = Pr$.

Example
A man has an annual income of $12,000 from two investments. He has $10,000 more invested at 8% than he has invested at 6%. How much does he have invested at each rate?

Solution
(a) Represent the amount invested at each rate symbolically.
Let $A$ = amount invested at 6%; then
$A + 10,000$ = amount invested at 8%.

Make a table.

| Investment rate | Amount | Interest |
|---|---|---|
| 6% or 0.06 | $A$ | $0.06A$ |
| 8% or 0.08 | $A + 10,000$ | $0.08(A + 10,000)$ |

(continued)

Write an equation relating the interest from each investment and the total interest received.

$$\begin{bmatrix} \text{Interest from} \\ 6\% \text{ investment} \end{bmatrix} + \begin{bmatrix} \text{interest from} \\ 8\% \text{ investment} \end{bmatrix} = [\text{total interest}]$$

$$0.06A \quad + 0.08(A + 10{,}000) = \quad 12{,}000$$

(b) Solve for $A$. First multiply each member by 100.

$$6A + 8(A + 10{,}000) = 1{,}200{,}000$$
$$6A + 8A + 80{,}000 = 1{,}200{,}000$$
$$14A = 1{,}120{,}000$$
$$A = 80{,}000;$$
$$A + 10{,}000 = 90{,}000$$

Therefore \$80,000 is invested at 6% and \$90,000 is invested at 8%.

25. A sum of \$2000 is invested, part at 6% and the remainder at 8%. Find the amount invested at each rate if the yearly income from the two investments is \$132.
26. A sum of \$2200 is invested, part at 5% and the remainder at 6%. Find the yearly interest on both investments if the interest on each investment is the same.
27. A man invests equal amounts of money at 5% and 7%. How much has he invested at each rate if his annual return is \$144?
28. A man has three times as much money invested in 8% bonds as he has in stocks paying 5%. How much does he have invested in each if his yearly income from the investments is \$1740?
29. A man has \$1000 more invested at 5% than he has invested at 8%. If his annual income from the two investments is \$739, how much does he have invested at each rate?
30. A man has \$500 more invested at 6% than he does at 8%. If his annual return on his investment is \$114, how much does he have invested at each rate?

*Uniform-motion Problems*

Solving uniform-motion problems requires applying the fact that the distance traveled at a uniform rate is equal to the product of the rate and the time traveled. This relationship may be expressed by any one of the three equations

$$d = rt, \quad r = \frac{d}{t}, \quad \text{and} \quad t = \frac{d}{r}$$

It is frequently helpful to use a table such as that shown in the following example.

Example    An express train travels 150 miles in the same time that a freight train travels 100 miles. If the express goes 20 miles per hour (mph) faster than the freight, find the rate of each.

## 1.5 Word Problems

Solution  (a) Represent the unknown rates in symbols.
Let $r$ = rate for freight train; then
$r + 20$ = rate for express train.

Make a table.

|         | d   | r      | t                  |
|---------|-----|--------|--------------------|
| Freight | 100 | $r$    | $\dfrac{100}{r}$   |
| Express | 150 | $r+20$ | $\dfrac{150}{r+20}$ |

Express the fact that the times are equal.

$$[t \text{ freight}] = [t \text{ express}]$$

$$\frac{100}{r} = \frac{150}{r+20}$$

(b) Solve the equation.

$$r(r+20)\frac{100}{r} = r(r+20)\frac{150}{r+20}$$
$$(r+20)100 = (r)150$$
$$100r + 2000 = 150r$$
$$-50r = -2000$$
$$r = 40;$$
$$r + 20 = 60$$

Thus, the freight train's rate is 40 mph and the express train's rate is 60 mph.

31. An airplane travels 1260 miles in the same time that an automobile travels 420 miles. If the rate of the airplane is 120 mph greater than the rate of the automobile, find the rate of each.

32. Two cars start together and travel in the same direction, one going twice as fast as the other. At the end of three hours they are 96 miles apart. How fast is each traveling?

33. A freight train leaves town A for town B, traveling at an average rate of 40 mph. Three hours later a passenger train also leaves town A for town B, traveling at an average rate of 80 mph. How far from town A does the passenger train pass the freight train?

34. Two planes leave an airport at the same time and travel in opposite directions. If one plane averages 440 mph over the ground and the other 520 mph, in how long will they be 2880 miles apart?

**35.** A car leaves a town at the rate of 42 mph. Two hours later a second car follows at the rate of 54 mph. How far from the town will they pass each other?

**36.** A boat sails due west from a harbor at 36 knots. An hour later, another boat leaves the harbor on the same course at 45 knots. How far from the harbor will the second boat overtake the first?

*Inequalities*

Sentences describing inequalities are set up in the same way as those describing equalities, except that symbols such as $\leq$ or $>$ replace the symbol $=$.

**Example**    A student must have an average of 80 to 90% on five tests in a course to receive a B. His grades on the first four tests were 98, 76, 86, and 92%. What grade on the fifth test would give him a B in the course?

**Solution**    (a) Represent the unknown quantity symbolically.

Let $x$ = grade (in percent) on the fifth test.

Write an inequality expressing the word sentence.

$$80 \leq \frac{98 + 76 + 86 + 92 + x}{5} < 90$$

(b) Solve. First multiply each member by 5.

$$400 \leq 352 + x < 450$$
$$48 \leq x < 98$$

Therefore, any grade equal to or greater than 48% and less than 98% would give the student a B.

**37.** In the preceding example, what grade on the fifth test would give the student a B if his grades on the first four tests were 78, 64, 88, and 76%?

**38.** The Fahrenheit and centigrade temperatures are related by the formula $C = \frac{5}{9}(F - 32)$. Within what range must the temperature be in Fahrenheit degrees for the temperature in centigrade degrees to lie between $-9$ and $27°$?

**39.** A man wishes to invest $10,000, part at 5% and part at 7%. What is the least he can invest at 7% if he wishes a minimum annual return of $616?

**40.** A man can sail upstream in a river at an average rate of 4 mph, and downstream at an average rate of 6 mph. What is the greatest distance he can sail upstream if he starts at 8 A.M. and must be back no later than 6 P.M.?

# UNIT REVIEW

[1.1–1.2]  1. Is $-3$ a solution of $\dfrac{2x+1}{5} - x = 4$?

2. Is $-3$ a solution of $4x - 2 = 6x + 4$?

*Solve each equation.*

3. $2(x - 3) = 12$
4. $3[x - 2(x + 1)] = 4 + x$
5. $\dfrac{x}{4} - 2 = \dfrac{x+1}{3}$
6. $\dfrac{2}{x-1} + \dfrac{1}{3(x-1)} = \dfrac{7}{2}$
7. $\dfrac{2t+6}{3} = \dfrac{1}{2} + t$
8. $\dfrac{1}{r} + \dfrac{1}{2r} = \dfrac{3}{2}$

[1.3]  *Solve each equation for the specified variable.*

9. $I = prt$, for $r$
10. $r = k + gt$, for $g$
11. $S = 3\pi d + 5\pi D$, for $D$
12. $\dfrac{1}{r} = \dfrac{1}{r_1} + \dfrac{1}{r_2}$, for $r$

[1.4]  *Solve each inequality and represent the solution set on a line graph.*

13. $\dfrac{2x}{3} \geq 4$
14. $-3x + 1 < 7$
15. $\dfrac{3x-2}{2} \leq 5$
16. $\dfrac{2}{5} + x > \dfrac{x}{2}$

[1.5]  17. A man has $20,000 to invest. He invests $5000 at 4% and $10,000 at 3%. At what rate should he invest the remainder in order to have a yearly income of $800?

18. A man has $1.60 in dimes and nickels, with two more nickels than dimes. How many of each coin does he have?

19. One freight train travels 6 mph faster than a second freight train. The first freight travels 280 miles in the same time the second freight travels 210 miles. Find the rate of each.

20. A man has $5000 invested at 4%. What is the least amount of money he can invest at 6% in order to receive at least 5% return on his total investment?

# UNIT 2

# SECOND-DEGREE EQUATIONS

In Unit 1 we considered first-degree equations and inequalities in one variable, and word problems which led to such mathematical models. In this unit we consider several methods of solving second-degree equations in one variable, and also word problems which lead to these models.

## 2.1 SOLUTION BY FACTORING

### Definitions

**Standard form for a second-degree equation**  An equation in the form  $ax^2 + bx + c = 0 \quad (a \neq 0)$

**Quadratic equation**  A second-degree equation

### Properties

If  $ab = 0$,  then  $a = 0$  or  $b = 0$.

### Exercises

What is the set of values of x for which each of the following products yields zero? Determine your answer by inspection or follow the procedure in the example.

Example  $(x - 3)(2x + 1)$

## 2.1 Solution by Factoring

**Solution**   If the product is zero, at least one factor is zero. So, set each factor equal to zero, and then solve the resulting equations.

$$x - 3 = 0; \quad 2x + 1 = 0$$

Solve each equation.

$$x = 3 \qquad x = -\frac{1}{2}$$

Hence, if $x$ is either 3 or $-\frac{1}{2}$, the original product will equal zero. The required set is $\left\{3, -\frac{1}{2}\right\}$.

1. $(x + 2)(x - 5)$
2. $(x + 3)(x - 4)$
3. $(2x + 5)(x - 2)$
4. $(x + 1)(3x - 1)$
5. $x(2x + 1)$
6. $x(3x - 7)$
7. $4(x - 6)(2x + 3)$
8. $5(2x - 7)(x + 1)$

*Solve. Determine your answer by inspection or follow the procedure in the example.*

**Example**   $x(x + 3) = 0$

**Solution**   Determine solutions by inspection or set each factor equal to zero and solve the equations.

$$x = 0; \quad x + 3 = 0$$
$$x = -3$$

Hence, if $x$ is either 0 or $-3$, the left-hand member of the original equation will equal zero. For $x = 0$,

$$0(0 + 3) = 0$$
$$0 \cdot 3 = 0$$

and for $x = -3$,

$$-3(-3 + 3) = 0$$
$$-3 \cdot 0 = 0$$

The solution set is $\{0, -3\}$.

9. $(x - 2)(2x + 1) = 0$
10. $(x + 5)(3x - 1) = 0$
11. $4(2x - 5)(3x + 2) = 0$
12. $2(3x - 4)(2x + 5) = 0$
13. $4x(2x - 9) = 0$
14. $3x(5x + 1) = 0$

**Example**   $x^2 + x = 30$

**Solution**   Write the equation in standard form so that the right-hand member is zero.

$$x^2 + x - 30 = 0$$

*(continued)*

Factor the left-hand member to obtain a product that is equal to zero.

$$(x + 6)(x - 5) = 0$$

Determine solutions by inspection or set each factor equal to zero and solve the equations.

$$x + 6 = 0; \quad x - 5 = 0$$
$$x = -6 \quad\quad x = 5$$

The solution set for the original quadratic equation is $\{-6, 5\}$.

15. $x^2 - 3x = 0$
16. $x^2 + 5x = 0$
17. $2x^2 = 6x$
18. $3x^2 = 3x$
19. $x^2 - 9 = 0$
20. $x^2 - 4 = 0$
21. $2x^2 - 18 = 0$
22. $3x^2 - 3 = 0$
23. $9x^2 = 4$
24. $4x^2 = 1$
25. $\frac{4}{9}x^2 - 1 = 0$
26. $\frac{9x^2}{25} - 1 = 0$
27. $x^2 - 5x + 4 = 0$
28. $x^2 + 5x + 6 = 0$
29. $x^2 - 5x - 14 = 0$
30. $x^2 - x - 42 = 0$
31. $3x^2 - 6x = -3$
32. $12x^2 = 8x + 15$

**Example**  $3x(x + 1) = 2x + 2$

**Solution**  Write the equation in standard form so that the right-hand member is zero.

$$3x^2 + x - 2 = 0$$

Factor the left-hand member to obtain a product that is equal to zero.

$$(3x - 2)(x + 1) = 0$$

Determine solutions by inspection or set each factor equal to zero and solve the equations.

$$3x - 2 = 0; \quad x + 1 = 0$$
$$x = \frac{2}{3} \quad\quad x = -1$$

The solution set is $\left\{\frac{2}{3}, -1\right\}$.

33. $x(2x - 3) = -1$
34. $2x(x - 2) = x + 3$
35. $(x - 2)(x + 1) = 4$
36. $x(3x + 2) = (x + 2)^2$

**Example**  $x^2 = \frac{13}{6}x - 1$

Solution    Multiply each member by the least common denominator, 6.

$$(6)x^2 = (6)\left(\frac{13}{6}x - 1\right)$$

$$(6)x^2 = (\cancel{6})\frac{13}{\cancel{6}}x - (\cancel{6})1$$

$$6x^2 = 13x - 6$$

Write the equation in standard form.

$$6x^2 - 13x + 6 = 0$$

Factor the left-hand member.

$$(2x - 3)(3x - 2) = 0$$

Determine the solutions by inspection or set each factor equal to 0 and solve the resulting equations.

$$2x - 3 = 0; \quad 3x - 2 = 0$$

$$2x = 3 \quad \quad 3x = 2$$

$$x = \frac{3}{2} \quad \quad x = \frac{2}{3}$$

The solution set is $\left\{\frac{3}{2}, \frac{2}{3}\right\}$.

37. $\dfrac{2x^2}{3} + \dfrac{x}{3} - 2 = 0$     38. $x - 1 = \dfrac{x^2}{4}$

39. $\dfrac{x^2}{6} + \dfrac{x}{3} = \dfrac{1}{2}$     40. $\dfrac{x}{4} - \dfrac{3}{4} = \dfrac{1}{x}$

41. $3 = \dfrac{10}{x^2} - \dfrac{7}{x}$     42. $\dfrac{4}{3x} + \dfrac{3}{3x+1} + 2 = 0$

43. $\dfrac{2}{x-3} - \dfrac{6}{x-8} = -1$     44. $\dfrac{x}{x-1} - \dfrac{x}{x+1} = \dfrac{4}{3}$

# 2.2  SOLUTIONS OF EQUATIONS OF THE FORM $x^2 = c$

## Property

If $c \geq 0$, then the solution set of $x^2 = c$ is $\{\sqrt{c}, -\sqrt{c}\}$.

# Unit 2 Second-degree Equations

## Exercises

*Solve each equation by using the property on page 25.*

**Example**  $7x^2 - 63 = 0$

**Solution**  Add 63 to each member, and then divide each member by 7, to obtain an equivalent equation with $x^2$ as the only term in the left-hand member.

$$7x^2 = 63$$
$$x^2 = 9$$

Set $x$ equal to each square root of 9.

$$x = 3; \quad x = -3$$

The solution set is $\{3, -3\}$.

1. $x^2 = 9$
2. $x^2 = 16$
3. $9x^2 = 25$
4. $4x^2 = 9$
5. $2x^2 = 14$
6. $3x^2 = 15$
7. $4x^2 - 24 = 0$
8. $3x^2 - 9 = 0$
9. $\dfrac{2x^2}{3} = 4$
10. $\dfrac{3x^2}{5} = 3$
11. $\dfrac{5x^2}{2} - 5 = 0$
12. $\dfrac{7x^2}{3} - 14 = 0$

**Example**  $(x + 3)^2 = 7$

**Solution**  Set $x + 3$ equal to each square root of 7 and solve for $x$.

$$x + 3 = \sqrt{7}; \quad x + 3 = -\sqrt{7}$$
$$x = -3 + \sqrt{7} \quad x = -3 - \sqrt{7}$$

The solution set is $\{-3 + \sqrt{7}, -3 - \sqrt{7}\}$.

13. $(x - 2)^2 = 9$
14. $(x + 3)^2 = 4$
15. $(2x - 1)^2 = 16$
16. $(3x + 1)^2 = 25$
17. $(x + 2)^2 = 3$
18. $(x - 5)^2 = 7$
19. $(x - 2)^2 = 12$
20. $(x + 3)^2 = 18$

## 2.3 SOLUTION BY COMPLETING THE SQUARE

### Definition

**Perfect square trinomial**  A trinomial of the form $x^2 + bx + \left(\dfrac{b}{2}\right)^2$, which is equal to $\left(x + \dfrac{b}{2}\right)^2$, the square of a binomial

## 2.3 Solution by Completing the Square

# Exercises

(a) What term must be added to each expression to make a perfect square?
(b) Write the resulting expression as the square of a binomial.

**Example** $x^2 - 9x$

**Solution** (a) Square one-half of $-9$, the coefficient of the first-degree term.

$$\left[\frac{1}{2}(-9)\right]^2 = \left(\frac{-9}{2}\right)^2 = \frac{81}{4}$$

(b) Rewrite $x^2 - 9x + \frac{81}{4}$ as the square of an expression.

$$\left(x - \frac{9}{2}\right)^2$$

1. $x^2 + 2x$
2. $x^2 - 4x$
3. $x^2 - 6x$
4. $x^2 + 10x$
5. $x^2 + 3x$
6. $x^2 - 5x$
7. $x^2 - 7x$
8. $x^2 + 11x$
9. $x^2 - x$
10. $x^2 + 15x$
11. $x^2 + \frac{1}{2}x$
12. $x^2 - \frac{3}{4}x$

*Solve each equation by completing the square.*

**Example** $2x^2 + 3x - 2 = 0$

**Solution** Rewrite the equation with the constant term as the right-hand member and the coefficient of $x^2$ equal to 1.

$$x^2 + \frac{3}{2}x = \frac{2}{2}$$

Add the square of one-half of the coefficient of the first-degree term to each member. That is, add $\left[\frac{1}{2}\left(\frac{3}{2}\right)\right]^2 = \frac{9}{16}$ to each member.

$$x^2 + \frac{3}{2}x + \frac{9}{16} = 1 + \frac{9}{16}$$

Rewrite the left-hand member as the square of an expression.

$$\left(x + \frac{3}{4}\right)^2 = \frac{25}{16}$$

*(continued)*

Set $x + \frac{3}{4}$ equal to each square root of $\frac{25}{16}$ and solve for $x$.

$$x + \frac{3}{4} = \frac{5}{4}; \quad x + \frac{3}{4} = -\frac{5}{4}$$

$$x = \frac{1}{2} \quad\quad x = -2$$

The solution set is $\left\{-2, \frac{1}{2}\right\}$.

13. $x^2 + 4x - 12 = 0$   14. $x^2 - x - 6 = 0$   15. $x^2 - 2x + 1 = 0$
16. $x^2 + 4x + 4 = 0$   17. $x^2 + 9x + 20 = 0$   18. $x^2 - x - 20 = 0$
19. $x^2 - 2x - 1 = 0$   20. $x^2 + 3x - 1 = 0$   21. $2x^2 = 4 - 3x$
22. $2x^2 = 6 - 5x$   23. $2x^2 + 4x = 3$   24. $3x^2 + x = 4$

## 2.4 SOLUTION BY THE QUADRATIC FORMULA

### Definition

**Quadratic formula**   $x = \dfrac{-b \pm \sqrt{b^2 - 4ac}}{2a}$ $(a \neq 0)$; the formula is equivalent to

$$x = \frac{-b + \sqrt{b^2 - 4ac}}{2a} \quad \text{or} \quad x = \frac{-b - \sqrt{b^2 - 4ac}}{2a}$$

### Property

For $a \neq 0$, the solution set of $ax^2 + bx + c = 0$ is

$$\left\{\frac{-b \pm \sqrt{b^2 - 4ac}}{2a}\right\}$$

### Exercises

Write each equation in standard form and specify the values of $a$, $b$, and $c$ for use in the quadratic formula.

## 2.4 Solution by the Quadratic Formula

Example $\quad x(2x+1) = 3$

Solution $\quad$ Apply the Distributive law to obtain
$$2x^2 + x = 3$$
and add $-3$ to both members to obtain
$$2x^2 + x - 3 = 0.$$
Comparing this equation to the standard form $ax^2 + bx + c = 0$, we have $a = 2$, $b = 1$, $c = -3$.

1. $x^2 - 3x = -4$
2. $x = 2 - 5x^2$
3. $x^2 + 2(x-1) = 0$
4. $3(x^2 - x) + 1 = 0$
5. $2x^2 - x = 0$
6. $x^2 = 3x$
7. $x(x-2) = 5$
8. $3x(x+1) = 4$
9. $(x+2)^2 = 6$
10. $(x-1)^2 = 7$
11. $(2x+1)^2 = x^2 - x$
12. $3x^2 - 2 = (x+3)^2$

*Solve each equation by using the quadratic formula.*

Example $\quad \dfrac{x^2}{4} - x = \dfrac{5}{4}$

Solution $\quad$ First multiply each member by the least common denominator, 4, and then write the equation in standard form.
$$x^2 - 4x = 5$$
$$x^2 - 4x - 5 = 0$$
Substitute 1 for $a$, $-4$ for $b$, and $-5$ for $c$ in the quadratic formula.
$$x = \frac{-(-4) \pm \sqrt{(-4)^2 - 4(1)(-5)}}{2(1)}$$
$$= \frac{4 \pm \sqrt{16 + 20}}{2}$$
$$= \frac{4 \pm 6}{2}$$
The solution set is $\{5, -1\}$.

13. $x^2 - 5x + 4 = 0$
14. $x^2 - 4x + 4 = 0$
15. $y^2 + 3y = 4$
16. $y^2 - 5y = 6$
17. $z^2 = 3z - 1$
18. $2z^2 = 7z - 2$
19. $0 = 3x^2 - 5x + 1$
20. $0 = 2x^2 - x - 2$
21. $5z + 6 = 6z^2$
22. $13z + 5 = 6z^2$
23. $x^2 + x = \dfrac{15}{4}$
24. $\dfrac{y^2}{2} - \dfrac{y}{2} = 1$
25. $\dfrac{x^2}{4} = 3 - \dfrac{x}{4}$
26. $\dfrac{x^2 - 3}{2} + \dfrac{x}{4} = 1$

## 2.5 WORD PROBLEMS

Quadratic equations are helpful in solving a variety of word problems. The suggestions on page 13 on how to solve word problems are also applicable to problems leading to quadratic equations.

## Exercises

*Solve.*

**Example**  The sum of twice the square of an integer and the integer itself is 15. Find the integer.

**Solution**  Let $x =$ an integer; then
$x^2 =$ the square of the integer.

The mathematical model is

$$2(x^2) + x = 15.$$

Writing the equation in standard form yields

$$2x^2 + x - 15 = 0$$

from which

$$(2x - 5)(x + 3) = 0.$$

By inspection or by setting each factor equal to zero, we obtain the solution set of the equation, $\left\{\frac{5}{2}, -3\right\}$. Since $\frac{5}{2}$ is not an integer, it *is not* a solution to the word problem. Since $2(-3)^2 + (-3) = 15$, the integer we want is $-3$.

1. Find two numbers whose sum is 13 and whose product is 42.
2. Find two numbers that differ by 3 and have a product of 40.
3. Find two consecutive positive odd integers whose product is 63.
4. Find two consecutive positive even integers whose product is 48.
5. The sum of a number and its reciprocal is $\frac{17}{4}$. What is the number?
6. The difference of a number and twice its reciprocal is $\frac{17}{3}$. What is the number?

**Example**  A ball thrown vertically upward reaches a height $h$ in feet, given by the equation $h = 64t - 16t^2$, where $t$ is the time in seconds after the throw. How long will it take the ball to reach a height of 48 feet on its way up?

## 2.5 Word Problems

**Solution**  To find $t$ when $h = 48$, we first substitute 48 for $h$ to obtain
$$48 = 64t - 16t^2$$
Writing the equation in standard form and then factoring the left-hand member, we obtain
$$16t^2 - 64t + 48 = 0$$
$$16(t^2 - 4t + 3) = 0$$
$$16(t - 1)(t - 3) = 0$$
The solution set of each equation is $\{1,3\}$. Hence, in 1 second the ball reaches a height of 48 feet on its way up. (In 3 seconds it reaches 48 feet on its way down.)

7. Use the mathematical model in the above example to find how long it will take the ball to reach a height of 64 feet.
8. How long after the throw will the ball in the above example return to the ground? (*Hint*: On the ground, $h = 0$.)
9. The distance $s$ a body falls in a vacuum is given by $s = v_0 t + \frac{1}{2}gt^2$, where $s$ is in feet, $t$ is in seconds, $v_0$ is the initial velocity in feet per second, and $g$ is the constant of acceleration due to gravity (approximately 32 feet per second per second). How long will it take a body to fall 150 feet if $v_0$ is 20 feet per second? How long if the body starts from rest?
10. How long will it take an object to fall 280 feet in a vacuum if it is given an initial velocity of 24 feet per second?

**Example**  A man rode a bicycle for 6 miles and then walked an additional 4 miles. The total time for his trip was 2 hours. If his rate walking was 6 mph less than his rate on the bicycle, what was each rate?

**Solution**  Let $x$ = rate when riding; then
$x - 6$ = rate when walking.

Since time = $\dfrac{\text{distance}}{\text{rate}}$, $\dfrac{6}{x}$ represents the *riding time* and $\dfrac{4}{x-6}$ represents the *walking time*. The total time of the trip is 6 hours. Hence

$$\text{Riding time} + \text{walking time} = \text{total time}$$
$$\frac{6}{x} + \frac{4}{x-6} = 2$$

from which
$$x(x-6)\left(\frac{6}{x} + \frac{4}{x-6}\right) = x(x-6)2, \quad (x \neq 0, 6)$$
$$6x - 36 + 4x = 2x^2 - 12x$$
$$2x^2 - 22x + 36 = 0$$
$$2(x - 2)(x - 9) = 0$$

The solution set is $\{2,9\}$.

(*continued*)

Although 2 is a solution of the equation, it *cannot be* the riding rate because the walking rate, which is 6 mph less than the riding rate, would be negative—an impossibility. The other solution, 9, does meet the conditions in the problem. Hence, the riding rate is 9 mph and the walking rate is 3 mph.

11. A man sailed a boat across a lake and back in $2\frac{1}{2}$ hours. If his rate returning was 2 mph less than his rate going, and if the distance each way was 6 miles, find his rate each way.
12. If a boat travels 20 mph in still water and takes 3 hours longer to go 90 miles up a river than it does to go 75 miles down the river, what is the speed of the current in the river?
13. A riverboat that travels at 18 mph in still water can go 30 miles up a river in 1 hour less time than it can go 63 miles down the river. What is the speed of the current in the river?
14. On a 50-mile trip, a man traveled 10 miles in traffic and then 40 miles in less congested driving. If his average rate in traffic was 20 mph less than his average rate out of traffic, what was each rate if the trip took 1 hour and 30 minutes?
15. A commuter takes a train for a trip of 10 miles to his job in a city. The train in the evening returns him home at a rate 10 mph greater than the rate of the train that takes him to work in the morning. If he spends a total 50 minutes a day commuting, what is the rate of each train?

# UNIT REVIEW

**[2.1]** Solve each equation by first writing the equation in standard form and then factoring the left-hand member.

1. $x^2 + 6 = 7x$
2. $x^2 = 4x$
3. $x(x+1) = 6$
4. $\dfrac{x^2}{4} + \dfrac{3x}{4} + \dfrac{1}{2} = 0$

**[2.2]** Solve each equation by using the fact that if $c \geq 0$, the solution set of $x^2 = c$ is $\{\sqrt{c}, -\sqrt{c}\}$.

5. $5x^2 = 20$
6. $2x^2 - 10 = 0$
7. $\dfrac{3x^2}{4} - 12 = 0$
8. $\dfrac{4x^2}{5} - 4 = 0$

**[2.3]** Write each of the following as the square of a binomial.

9. $x^2 + 6x + 9$
10. $x^2 - 8x + 16$
11. $x^2 + 3x + \dfrac{9}{4}$
12. $x^2 - x + \dfrac{1}{4}$

Solve each equation by completing the square.

13. $x^2 - 6x - 7 = 0$
14. $2x^2 + 5x - 3 = 0$

**[2.4]** Solve each equation by using the quadratic formula.

15. $x^2 + 3x - 10 = 0$
16. $y^2 - 2y = 8$
17. $\dfrac{y^2}{4} + y + \dfrac{3}{4} = 0$
18. $x^2 + \dfrac{3}{2}x = \dfrac{1}{2}$

**[2.5]**
19. The sum of a number and two times its reciprocal is $\tfrac{73}{6}$. What is the number?
20. A man drives his car for 20 miles at a certain speed. He then increases his speed by 20 mph and continues for an additional 30 miles. If the total trip takes 1 hour, how fast was the man traveling originally?

# APPENDIX

# QUADRATIC INEQUALITIES

## Definition

**Sign graph**  One or more number lines (used to visualize the solution set of a quadratic inequality) which indicate the signs associated with each factor of a product for number replacements of the variable

## Exercises

*Solve each of the following inequalities in two ways:*
*(a) By algebraic methods.*
*(b) By using a sign graph.*

Example $\quad x^2 + 4x < 5$

Solution $\quad$ (a) We first rewrite the inequality equivalently as
$$x^2 + 4x - 5 < 0$$
and then as
$$(x + 5)(x - 1) < 0$$
Only those values of $x$ for which the factors $x + 5$ and $x - 1$ are opposite in sign will be in the solution set. These can be determined analytically by noting that $(x + 5)(x - 1) < 0$ implies either
$$x + 5 < 0 \quad \text{and} \quad x - 1 > 0$$
or else
$$x + 5 > 0 \quad \text{and} \quad x - 1 < 0$$

Each of these two cases can be considered separately. First,
$$x+5<0 \quad \text{and} \quad x-1>0$$
imply
$$x<-5 \quad \text{and} \quad x>1$$
a condition which is not satisfied by any values of $x$. The inequalities
$$x+5>0 \quad \text{and} \quad x-1<0$$
imply
$$x>-5 \quad \text{and} \quad x<1$$
which lead to the solution set
$$S=\{x|-5<x<1\}$$

(b) The sign graph is constructed by first showing on a line the places where $x+5$ is positive ($x>-5$) and the places where it is negative ($x<-5$), and then showing on a second line those places where $x-1$ is positive ($x>1$) and those where it is negative ($x<1$). A number line can then be marked by observing those parts of the first two lines where the signs are alike and those parts where the signs are opposite. Since it is desired that the product $(x+5)(x-1)$ be negative, the number line shows clearly that this occurs where $-5<x<1$, so that the solution set of the inequality is
$$\{x|-5<x<1\}$$

```
(x + 5) - - o + + + + + + + + + +
(x - 1) - - - - - - - - - o + + +
         |  |  |  |  |  |  |  |  |  |  |
        -5              0              5
```
(Opposite Signs Excluding Endpoints)
$$\{x|x^2+4x-5<0\}=\{x|-5<x<1\}$$

1. $(x+1)(x-2)>0$
2. $(x+2)(x+5)<0$
3. $x(x-2)\leq 0$
4. $x(x+3)\geq 0$
5. $x^2-3x-4>0$
6. $x^2-5x-6\geq 0$

*Solve by algebraic methods.*

**Example**
$$\frac{x}{x-2}\geq 5$$

**Solution** If each member of such an inequality is multiplied by an expression containing the variable, we have to be careful either to distinguish between those values of the variable for which the expression denotes a positive and negative number, respectively, or to make sure that the expression

*(continued)*

by which we multiply is always positive. In this example the necessity of considering separate cases for $x - 2 > 0$ and $x - 2 < 0$ can be avoided if the given inequality is "cleared" of fractions by multiplying each member by $(x - 2)^2$, which is positive for $x \neq 2$. Doing this we have

$$(x - 2)^2 \frac{x}{(x - 2)} \geq (x - 2)^2 \cdot 5$$
$$x^2 - 2x \geq 5x^2 - 20x + 20$$
$$0 \geq 4x^2 - 18x + 20$$
$$0 \geq 2x^2 - 9x + 10$$
$$0 \geq (2x - 5)(x - 2)$$

This latter inequality can then be solved by noting that for $(2x - 5)(x - 2)$ to be nonpositive, either

$$(2x - 5) \geq 0 \quad \text{and} \quad (x - 2) < 0$$

or

$$(2x - 5) \leq 0 \quad \text{and} \quad (x - 2) > 0$$

must hold. The first of these implies that

$$x \geq \frac{5}{2} \quad \text{and} \quad x < 2$$

which have no common elements in their solution sets, while the second implies that

$$x \leq \frac{5}{2} \quad \text{and} \quad x > 2$$

which leads to the solution set $\left\{ x \mid 2 < x \leq \frac{5}{2} \right\}$.

7. $\dfrac{2}{x} \leq 4$     8. $\dfrac{3}{x - 6} > 8$     9. $\dfrac{x}{x + 2} > 4$     10. $\dfrac{x + 2}{x - 2} \geq 6$

# SOLUTION KEY

## [1.1]

1. Does $(4) - 3 = 7$?  No.

2. Does $2(4) - 6 = 3$?  No.

3. Does $2(3) + 6 = 3(3) - 5$?   Does $6 + 6 = 9 - 5$?  No.

4. Does $5(1) - 1 = 2(1) + 2$?   Does $5 - 1 = 2 + 2$?  Yes.

5. Does $3(3) + 2 = 8 + (3)$?   Does $9 + 2 = 8 + 3$?  Yes.

6. Does $3(-1) - 5 = 2(-1) + 7$?   Does $-3 - 5 = -2 + 7$?  No.

7. Does $3 = 6(-1) + 3[-1 - 2]$?   Does $3 = -6 - 9$?  No.

8. Does $6 - 2(0) = 6(2 \cdot 0 + 1)$?   Does $6 - 0 = 6(1)$?  Yes.

9. Does $0 = 6(4) - 24$?   Does $0 = 24 - 24$?  Yes.

10. Does $0 = 7(-1) + 7$?   Does $0 = -7 + 7$?  Yes.

11. Does $\frac{1}{4}(8) - 3 = 8 + 2$?   Does $2 - 3 = 10$?  No.

12. Does $\frac{1}{3}(6) - 2(6) = -12 + \frac{1}{3}(6)$?   Does $2 - 12 = -12 + 2$?  Yes.

**Solution Key 1.1**

**13.** $x + 2 = 5$
Subtract 2.
$x + 2 - 2 = 5 - 2$
$x = 3$
Solution set: {3}

**14.** $x - 3 = 7$
Add 3.
$x - 3 + 3 = 7 + 3$
$x = 10$
Solution set: {10}

**15.** $12 = x + 7$
Subtract 7.
$12 - 7 = x + 7 - 7$
$5 = x$
Solution set: {5}

**16.** $6 = x - 3$
Add 3.
$6 + 3 = x - 3 + 3$
$9 = x$
Solution set: {9}

**17.** $2x = 8$
Divide by 2.
$\dfrac{2x}{2} = \dfrac{8}{2}$
$x = 4$
Solution set: {4}

**18.** $7x = 21$
Divide by 7.
$\dfrac{7x}{7} = \dfrac{21}{7}$
$x = 3$
Solution set: {3}

**19.** $-12 = -3x$
Divide by $-3$.
$\dfrac{-12}{-3} = \dfrac{-3x}{-3}$
$4 = x$
Solution set: {4}

**20.** $-6 = 2x$
Divide by 2.
$\dfrac{-6}{2} = \dfrac{2x}{2}$
$-3 = x$
Solution set: {$-3$}

**21.** $-\dfrac{x}{3} = 5$
Multiply by $-3$.
$-3\left(-\dfrac{x}{3}\right) = (-3)5$
$x = -15$
Solution set: {$-15$}

**22.** $\dfrac{x}{6} = 2$
$(6)\dfrac{x}{6} = (6)2$
$x = 12$
Solution set: {12}

**23.** $\dfrac{x}{4} = -1$
$4\left(\dfrac{x}{4}\right) = 4(-1)$
$x = -4$
Solution set: {$-4$}

**24.** $\dfrac{x}{-6} = -3$
$-6\left(\dfrac{x}{-6}\right) = -6(-3)$
$x = 18$
Solution set: {18}

**25.** $x = 3 - 2x$
$x + 2x = 3 - 2x + 2x$
$3x = 3$
$\dfrac{3x}{3} = \dfrac{3}{3}$
$x = 1$
Solution set: {1}

**26.** $2x = 3 + x$
$2x - x = 3 + x - x$
$x = 3$
Solution set: {3}

Solution Key 1.1

**27.**
$$4 - x = x + 2$$
Add $x - 2$.
$$4 - x + x - 2 = x + 2 + x - 2$$
$$2 = 2x$$
$$\frac{2}{2} = \frac{2x}{2}$$
$$1 = x$$
Solution set: $\{1\}$

**28.**
$$2 - x = 4 + 3x$$
Add $x - 4$.
$$2 - x + x - 4 = 4 + 3x + x - 4$$
$$-2 = 4x$$
$$\frac{-2}{4} = \frac{4x}{4}$$
$$-\frac{1}{2} = x$$
Solution set: $\left\{-\frac{1}{2}\right\}$

**29.**
$$3 = x - 4 + 6x$$
$$3 + 4 = x - 4 + 6x + 4$$
$$7 = 7x$$
$$\frac{7}{7} = \frac{7x}{7}$$
$$1 = x$$
Solution set: $\{1\}$

**30.**
$$3x + 2 = 4x - 6 + x$$
$$3x + 2 - 3x + 6 = 4x - 6 + x - 3x + 6$$
$$8 = 2x$$
$$\frac{8}{2} = \frac{2x}{2}$$
$$4 = x$$
Solution set: $\{4\}$

**31.** $E = 2.22, \quad R = 0.60$
$$E = IR$$
Substituting,
$$2.22 = I(0.60)$$
$$\frac{2.22}{0.60} = \frac{I(0.60)}{0.60}$$
$$3.70 = I$$

**32.** $C = 42.31, \quad \pi = 3.14$
$$C = \pi D$$
Substituting,
$$42.31 = 3.14 D$$
$$\frac{42.31}{3.14} = \frac{3.14 D}{3.14}$$
$$13.47 = D$$

**33.** $P = 42, \quad l = 6$
$$P = 2l + 2w$$
Substituting,
$$42 = 2(6) + 2w$$
$$42 = 12 + 2w$$
$$30 = 2w$$
$$15 = w$$

**34.** $A = 642, \quad r = 0.04$
$$A = P(1 + r)$$
Substituting,
$$642 = P(1 + 0.04)$$
$$642 = P(1.04)$$
$$\frac{642}{1.04} = P$$
$$617.31 = P$$

**35.** $A = 46.12, \quad r = 2.00, \quad \pi = 3.14$
$$A = \pi r^2 h$$
$$46.12 = (3.14)(2.00)^2 h$$
$$46.12 = (3.14)(4)h$$
$$46.12 = 12.56 h$$
$$\frac{46.12}{12.56} = h$$
$$3.67 = h$$

**36.** $A = 235.5, \quad r = 3.0, \quad \pi = 3.14$
$$A = \pi r^2 + 2\pi r h$$
$$235.5 = 3.14(3.0)^2 + 2(3.14)(3.0)h$$
$$235.5 = 28.26 + 18.84 h$$
$$235.5 - 28.26 = 28.26 + 18.84 h - 28.26$$
$$207.24 = 18.84 h$$
$$\frac{207.24}{18.84} = \frac{18.84 h}{18.84}$$
$$11.0 = h$$

# [1.2]

**1.**
$$2(x+5) = 16$$
$$2x + 10 = 16$$
$$2x + 10 - 10 = 16 - 10$$
$$2x = 6$$
$$\frac{2x}{3} = \frac{6}{2}$$
$$x = 3$$
Solution set: {3}

**2.**
$$3(x+1) = 3$$
$$3x + 3 = 3$$
$$3x + 3 - 3 = 3 - 3$$
$$3x = 0$$
$$\frac{3x}{3} = \frac{0}{3}$$
$$x = 0$$
Solution set: {0}

**3.**
$$6 = 2(2x - 1)$$
$$6 = 4x - 2$$
$$6 + 2 = 4x - 2 + 2$$
$$8 = 4x$$
$$\frac{8}{4} = \frac{4x}{4}$$
$$2 = x$$
Solution set: {2}

**4.**
$$3 = 4(2x + 1)$$
$$3 = 8x + 4$$
$$3 - 4 = 8x + 4 - 4$$
$$-1 = 8x$$
$$\frac{-1}{8} = \frac{8x}{8}$$
$$\frac{-1}{8} = x$$
Solution set: $\left\{-\frac{1}{8}\right\}$

**5.**
$$x - (8 - x) = 2$$
$$x - 8 + x = 2$$
$$2x - 8 = 2$$
$$2x - 8 + 8 = 2 + 8$$
$$2x = 10$$
$$\frac{2x}{2} = \frac{10}{2}$$
$$x = 5$$
Solution set: {5}

**6.**
$$2x - (3 - x) = 0$$
$$2x - 3 + x = 0$$
$$3x - 3 = 0$$
$$3x - 3 + 3 = 0 + 3$$
$$3x = 3$$
$$\frac{3x}{3} = \frac{3}{3}$$
$$x = 1$$
Solution set: {1}

**7.**
$$3(7 - 2x) = 30 - 7(x + 1)$$
$$21 - 6x = 30 - 7x - 7$$
$$21 - 6x = 23 - 7x$$
$$7x - 6x = 23 - 21$$
$$x = 2$$
Solution set: {2}

**8.**
$$6x - 2(2x + 5) = 6(5 + x)$$
$$6x - 4x - 10 = 30 + 6x$$
$$2x - 10 = 30 + 6x$$
$$-10 - 30 = 6x - 2x$$
$$-40 = 4x$$
$$-10 = x$$
Solution set: {−10}

**9.**
$$-3[x - (2x + 3) - 2x] = -9$$
$$-3[x - 2x - 3 - 2x] = -9$$
$$-3[-3x - 3] = -9$$
$$9x + 9 = -9$$
$$9x = -9 - 9$$
$$9x = -18$$
$$x = -2$$
Solution set: {−2}

**10.**
$$5[2 + 3(x - 2)] + 20 = 0$$
$$5[2 + 3x - 6] + 20 = 0$$
$$5[-4 + 3x] + 20 = 0$$
$$-20 + 15x + 20 = 0$$
$$15x = 0$$
$$x = 0$$
Solution set: {0}

Solution Key  1.2

**11.**  $-2[x - (x - 1)] = -3(x + 1)$
$-2[x - x + 1] = -3x - 3$
$-2[1] = -3x - 3$
$-2 + 3 = -3x$
$1 = -3x$
$\dfrac{-1}{3} = x$
Solution set: $\left\{\dfrac{-1}{3}\right\}$

**12.**  $3[2x - (x + 2)] = -3(3 - 2x)$
$3[2x - x - 2] = -9 + 6x$
$3[x - 2] = -9 + 6x$
$3x - 6 = -9 + 6x$
$-6 + 9 = 6x - 3x$
$3 = 3x$
$1 = x$
Solution set: $\{1\}$

**13.**  $(x - 1)(x - 1) = x^2 - 11$
$x^2 - 2x + 1 = x^2 - 11$
$x^2 - 2x - x^2 = -11 - 1$
$-2x = -12$
$x = 6$
Solution set: $\{6\}$

**14.**  $(2x + 1)(x - 3) = (x - 2)(2x + 1)$
$2x^2 - 5x - 3 = 2x^2 - 3x - 2$
$2x^2 - 5x - 2x^2 + 3x = -2 + 3$
$-2x = 1$
$x = \dfrac{-1}{2}$
Solution set: $\left\{\dfrac{-1}{2}\right\}$

**15.**  $(x - 2)^2 = x^2 - 8$
$(x - 2)(x - 2) = x^2 - 8$
$x^2 - 4x + 4 = x^2 - 8$
$x^2 - 4x - x^2 = -8 - 4$
$-4x = -12$
$x = 3$
Solution set: $\{3\}$

**16.**  $(x - 1)^2 = x^2 - 15$
$(x - 1)(x - 1) = x^2 - 15$
$x^2 - 2x + 1 = x^2 - 15$
$x^2 - 2x - x^2 = -15 - 1$
$-2x = -16$
$x = 8$
Solution set: $\{8\}$

**17.**  $4 - (x - 3)(x + 2) = 10 - x^2$
$4 - (x^2 - x - 6) = 10 - x^2$
$4 - x^2 + x + 6 = 10 - x^2$
$10 + x = 10$
$x = 10 - 10$
$x = 0$
Solution set: $\{0\}$

**18.**  $0 = x^2 + (2 - x)(5 + x)$
$0 = x^2 + 10 - 3x - x^2$
$0 = 10 - 3x$
$3x = 10$
$x = \dfrac{10}{3}$
Solution set: $\left\{\dfrac{10}{3}\right\}$

**19.**  $\dfrac{x}{3} = 2$
$3\left(\dfrac{x}{3}\right) = 3(2)$
$x = 6$
Solution set: $\{6\}$

**20.**  $\dfrac{2x}{3} = -4$
$3\left(\dfrac{2x}{3}\right) = 3(-4)$
$2x = -12$
$x = -6$
Solution set: $\{-6\}$

**21.** $\dfrac{5x - x}{6} = 4$

$6\left(\dfrac{5x - x}{6}\right) = 6(4)$

$5x - x = 24$

$4x = 24$

$x = 6$

Solution set: {6}

**22.** $7 - \dfrac{x}{3} = x - 1$

$3\left(7 - \dfrac{x}{3}\right) = 3(x - 1)$

$21 - 3\left(\dfrac{x}{3}\right) = 3x - 3$

$21 - x = 3x - 3$

$21 + 3 = 3x + x$

$24 = 4x$

$6 = x$

Solution set: {6}

**23.** $6 - \dfrac{x}{2} = x$

$2\left(6 - \dfrac{x}{2}\right) = 2x$

$12 - 2\left(\dfrac{x}{2}\right) = 2x$

$12 - x = 2x$

$12 = 2x + x$

$12 = 3x$

$4 = x$

Solution set: {4}

**24.** $\dfrac{2x}{3} - 3 = x - 5$

$3\left(\dfrac{2x}{3} - 3\right) = 3(x - 5)$

$3\left(\dfrac{2x}{3}\right) - 9 = 3x - 15$

$2x - 9 = 3x - 15$

$-9 + 15 = 3x - 2x$

$6 = x$

Solution set: {6}

**25.** $2 + \dfrac{x}{6} = \dfrac{5}{2}$

Multiply by the least common denominator, 6.

$6\left(2 + \dfrac{x}{6}\right) = 6\left(\dfrac{5}{2}\right)$

$12 + x = 15$

$12 + x - 12 = 15 - 12$

$x = 3$

Solution set: {3}

**26.** $1 + \dfrac{x}{9} = \dfrac{4}{3}$

Multiply by the least common denominator, 9.

$9\left(1 + \dfrac{x}{9}\right) = 9\left(\dfrac{4}{3}\right)$

$9 + x = 12$

$x = 12 - 9$

$x = 3$

Solution set: {3}

**27.** $\dfrac{x}{5} - \dfrac{x}{2} = 9$

Multiply by the least common denominator, 10.

$10\left(\dfrac{x}{5} - \dfrac{x}{2}\right) = 10(9)$

$2x - 5x = 90$

$-3x = 90$

$x = -30$

Solution set: {−30}

**28.** $\dfrac{y}{2} + \dfrac{y}{3} - \dfrac{y}{4} = 7$

Multiply by the least common denominator, 12.

$12\left(\dfrac{y}{2} + \dfrac{y}{3} - \dfrac{y}{4}\right) = 12(7)$

$6y + 4y - 3y = 84$

$7y = 84$

$y = 12$

Solution set: {12}

**29.**
$$\frac{2x-1}{5} = \frac{x+1}{2}$$
$$10\left(\frac{2x-1}{5}\right) = 10\left(\frac{x+1}{2}\right)$$
$$4x - 2 = 5x + 5$$
$$-2 - 5 = 5x - 4x$$
$$-7 = x$$
Solution set: $\{-7\}$

**30.**
$$\frac{2x}{3} - \frac{2x+5}{6} = \frac{1}{2}$$
$$6\left(\frac{2x}{3} - \frac{2x+5}{6}\right) = 6\left(\frac{1}{2}\right)$$
$$4x - (2x + 5) = 3$$
$$4x - 2x - 5 = 3$$
$$2x = 3 + 5$$
$$x = 4$$
Solution set: $\{4\}$

**31.**
$$\frac{3}{5} = \frac{x}{x+2}$$
Multiply by the least common denominator, $5(x+2)$.
$$5(x+2)\left(\frac{3}{5}\right) = 5(x+2)\left(\frac{x}{x+2}\right)$$
$$3(x+2) = 5x$$
$$3x + 6 = 5x$$
$$6 = 5x + (-3x)$$
$$6 = 2x$$
$$3 = x$$
Solution set: $\{3\}$

**32.**
$$\frac{2}{x-2} = \frac{9}{x+12}$$
Multiply by the least common denominator, $(x-2)(x+12)$.
$$(x-2)(x+12)\left(\frac{2}{x-2}\right) = (x-2)(x+12)\left(\frac{9}{x+12}\right)$$
$$2(x+12) = 9(x-2)$$
$$2x + 24 = 9x - 18$$
$$2x + 24 - 2x + 18 = 9x - 18 - 2x + 18$$
$$42 = 7x$$
$$6 = x$$
Solution set: $\{6\}$

**33.**
$$\frac{x}{x-2} = \frac{2}{x-2} + 2$$
$$(x-2)\left(\frac{x}{x-2}\right) = (x-2)\left(\frac{2}{x-2} + 2\right)$$
$$(x-2)\left(\frac{x}{x-2}\right) = (x-2)\left(\frac{2}{x-2}\right) + (x-2)2$$
$$x = 2 + 2(x-2)$$
$$x = 2 + 2x - 4$$
$$x = 2x - 2$$
$$2 = 2x - x$$
$$2 = x$$
But since 2 is not a solution of the original equation, there is no solution to the original equation.

**34.**
$$\frac{5}{x-3} = \frac{x+2}{x-3} + 2$$
$$(x-3)\left(\frac{5}{x-3}\right) = (x-3)\left(\frac{x+2}{x-3} + 2\right)$$
$$(x-3)\left(\frac{5}{x-3}\right) = (x-3)\left(\frac{x+2}{x-3}\right) + (x-3)2$$
$$5 = x + 2 + 2(x-3)$$
$$5 = x + 2 + 2x - 6$$
$$5 = 3x - 4$$
$$9 = 3x$$
$$3 = x$$

But since 3 is not a solution of the original equation, there is no solution to the original equation.

**35.**
$$\frac{2}{y+1} + \frac{1}{3y+3} = \frac{1}{6}$$
Factor the denominator, $3y+3$.
$$\frac{2}{y+1} + \frac{1}{3(y+1)} = \frac{1}{6}$$
Multiply by the least common denominator, $6(y+1)$.
$$6(y+1)\left[\frac{2}{y+1} + \frac{1}{3(y+1)}\right] = 6(y+1)\frac{1}{6}$$
$$12 + 2 = y + 1$$
$$14 = y + 1$$
$$13 = y$$

Solution set: $\{13\}$

**36.**
$$\frac{y}{y+2} - \frac{3}{y-2} = \frac{y^2+8}{y^2-4}$$
Multiply by the least common denominator, $(y+2)(y-2)$. [Note: $(y+2)(y-2) = y^2 - 4$]
$$(y+2)(y-2)\left(\frac{y}{y+2} - \frac{3}{y-2}\right) = (y+2)(y-2)\left(\frac{y^2+8}{y^2-4}\right)$$
$$y(y-2) - 3(y+2) = y^2 + 8$$
$$y^2 - 2y - 3y - 6 = y^2 + 8$$
$$y^2 - 5y - 6 = y^2 + 8$$
$$-5y = 8 + 6$$
$$-5y = 14$$
$$y = \frac{-14}{5}$$

Solution set: $\left\{\frac{-14}{5}\right\}$

**37.** $F = 176$
$$F = \frac{9}{5}C + 32$$
Substituting,
$$176 = \frac{9}{5}C + 32$$
$$144 = \frac{9}{5}C$$
$$\frac{5}{9}(144) = \frac{5}{9}\left(\frac{9}{5}C\right)$$
$$80 = C$$

**38.** $g = 32.2, \quad t = 4$
$$s - \frac{1}{2}gt^2 = 0$$
Substituting,
$$s - \frac{1}{2}(32.2)(4^2) = 0$$
$$s - 257.6 = 0$$
$$s = 257.6$$

**Solution Key 1.3**

39. $a = 6,\ S = 9$
$$S = \frac{a}{1-r}$$
Substituting,
$$9 = \frac{6}{1-r}$$
$$(1-r)9 = (1-r)\left(\frac{6}{1-r}\right)$$
$$9 - 9r = 6$$
$$-9r = -3$$
$$r = \frac{1}{3}$$

40. $A = 240,\ h = 20,\ c = 12$
$$A = \frac{h}{2}(b+c)$$
Substituting,
$$240 = \frac{20}{2}(b+12)$$
$$240 = 10(b+12)$$
$$240 = 10b + 120$$
$$240 + (-120) = 10b$$
$$120 = 10b$$
$$12 = b$$

**[1.3]**

1. $ax = b + c$
Divide by $a$.
$$\frac{ax}{a} = \frac{b+c}{a}$$
$$x = \frac{b+c}{a}$$

2. $b - c = ay$
Divide by $a$.
$$\frac{b-c}{a} = \frac{ay}{a}$$
$$\frac{b-c}{a} = y$$
or $y = \frac{b-c}{a}$

3. $3ay - b = c$
Add $b$.
$$3ay - b + b = c + b$$
$$3ay = c + b$$
Divide by $3a$.
$$\frac{3ay}{3a} = \frac{c+b}{3a}$$
$$y = \frac{c+b}{3a}$$

4. $4by + c = b$
Subtract $c$.
$$4by + c - c = b - c$$
$$4by = b - c$$
Divide by $4b$.
$$\frac{4by}{4b} = \frac{b-c}{4b}$$
$$y = \frac{b-c}{4b}$$

5. $ax = b - 2ax$
Add $2ax$.
$$ax + 2ax = b - 2ax + 2ax$$
$$3ax = b$$
Divide by $3a$.
$$\frac{3ax}{3a} = \frac{b}{3a}$$
$$x = \frac{b}{3a}$$

6. $4by = a + 2by$
Subtract $2by$.
$$4by - 2by = a + 2by - 2by$$
$$2by = a$$
Divide by $2b$.
$$\frac{2by}{2b} = \frac{a}{2b}$$
$$y = \frac{a}{2b}$$

7. $\frac{a}{b}y + c = 0$
$$\frac{a}{b}y = -c$$
$$\left(\frac{b}{a}\right)\left(\frac{a}{b}\right)y = \left(\frac{b}{a}\right)(-c)$$
$$y = \frac{-bc}{a}$$

8. $b = \frac{1}{c}y + a$
$$b - a = \frac{1}{c}y$$
$$(c)(b-a) = (c)\left(\frac{1}{c}\right)y$$
$$c(b-a) = y$$
or $y = cb - ca$

## Solution Key 1.3

**9.** $\dfrac{x+a}{b} = c$

(b)$\dfrac{x+a}{b} = (b)c$

$x + a = bc$

$x = bc - a$

**10.** $b = \dfrac{y-a}{c}$

$b(c) = \dfrac{y-a}{c}(c)$

$bc = y - a$

$bc + a = y$

$y = bc + a$

**11.** $\dfrac{c-2y}{b} = a$

$\dfrac{c-2y}{b}(b) = a(b)$

$c - 2y = ab$

$-2y = ab - c$

$y = \dfrac{ab - c}{-2}$

or $y = \dfrac{c - ab}{2}$

**12.** $\dfrac{b-3x}{a} = c$

(a)$\dfrac{b-3x}{a} = (a)c$

$b - 3x = ac$

$-3x = ac - b$

$x = \dfrac{ac - b}{-3}$

or $x = \dfrac{b - ac}{3}$

**13.** $ax = a - x$

$ax + x = a$

$x(a + 1) = a$

$x = \dfrac{a}{a+1}$

**14.** $y = b + by$

$y - by = b$

$(1 - b)y = b$

$y = \dfrac{b}{1 - b}$

**15.** $a(a - x) = b(b - x)$

$a^2 - ax = b^2 - bx$

$-ax + bx = b^2 - a^2$

$bx - ax = b^2 - a^2$

$(b - a)x = (b - a)(b + a)$

$x = b + a$

**16.** $4y - 3(y - b) = 8b$

$4y - 3y + 3b = 8b$

$y + 3b = 8b$

$y = 5b$

**17.** $(x - 4)(b + 2) = b$

$\dfrac{(x-4)(b+2)}{b+2} = \dfrac{b}{b+2}$

$x - 4 = \dfrac{b}{b+2}$

$x = \dfrac{b}{b+2} + 4$

or $x = \dfrac{b + 4(b+2)}{b+2}$

$x = \dfrac{5b + 8}{b+2}$

**18.** $(y - 2)(a + 3) = 2a$

$\dfrac{(y-2)(a+3)}{a+3} = \dfrac{2a}{a+3}$

$y - 2 = \dfrac{2a}{a+3}$

$y = \dfrac{2a}{a+3} + 2$

or $y = \dfrac{2a + 2(a+3)}{a+3}$

$y = \dfrac{4a + 6}{a+3}$

## Solution Key 1.3

**19.** $\dfrac{2}{x} + \dfrac{2}{a} = 6$

Multiply by the least common denominator, $ax$.

$(ax)\dfrac{2}{x} + (ax)\dfrac{2}{a} = 6(ax)$

$2a + 2x = 6ax$

$2a = 6ax - 2x$

$2a = 2(3a - 1)x$

$\dfrac{2a}{2(3a - 1)} = x$

$x = \dfrac{a}{3a - 1}$

**20.** $\dfrac{2}{y} - \dfrac{3}{b} = 4$

Multiply by the least common denominator, $by$.

$(by)\dfrac{2}{y} - (by)\dfrac{3}{b} = 4(by)$

$2b - 3y = 4by$

$2b = 4by + 3y$

$2b = (4b + 3)y$

$\dfrac{2b}{4b + 3} = y$

$y = \dfrac{2b}{4b + 3}$

**21.** $\dfrac{a}{y} - \dfrac{1}{2} = \dfrac{a}{2y}$

Multiply by the least common denominator, $2y$.

$(2y)\dfrac{a}{y} - (2y)\dfrac{1}{2} = (2y)\dfrac{a}{2y}$

$2a - y = a$

$2a - a = y$

$y = a$

**22.** $\dfrac{c}{3} + \dfrac{1}{x} = \dfrac{c}{6x}$

Multiply by the least common denominator, $6x$.

$(6x)\dfrac{c}{3} + (6x)\dfrac{1}{x} = (6x)\dfrac{c}{6x}$

$2cx + 6 = c$

$2cx = c - 6$

$x = \dfrac{c - 6}{2c}$

**23.** $\dfrac{1}{a} + \dfrac{1}{b} = \dfrac{1}{x}$

$(abx)\dfrac{1}{a} + (abx)\dfrac{1}{b} = (abx)\dfrac{1}{x}$

$bx + ax = ab$

$(b + a)x = ab$

$x = \dfrac{ab}{b + a}$

**24.** $\dfrac{1}{a} - \dfrac{1}{b} = \dfrac{1}{x}$

$(abx)\dfrac{1}{a} - (abx)\dfrac{1}{b} = (abx)\dfrac{1}{x}$

$bx - ax = ab$

$(b - a)x = ab$

$x = \dfrac{ab}{b - a}$

**25.** $v = k + gt$

$v - gt = k + gt - gt$

$v - gt = k$

$k = v - gt$

**26.** $E = mc^2$

$\dfrac{E}{c^2} = m$

$m = \dfrac{E}{c^2}$

**27.** $f = ma$

$\dfrac{f}{a} = m$

$m = \dfrac{f}{a}$

**28.** $I = prt$

$\dfrac{I}{rt} = p$

$p = \dfrac{I}{rt}$

**29.** $pv = K$

$v = \dfrac{K}{p}$

**30.** $E = IR$

$\dfrac{E}{I} = R$

$R = \dfrac{E}{I}$

**31.** $s = \dfrac{1}{2}at^2$

$2s = at^2$

$\dfrac{2s}{t^2} = a$

$a = \dfrac{2s}{t^2}$

**32.** $p = 2l + 2w$

$p - 2w = 2l$

$\dfrac{p - 2w}{2} = l$

$l = \dfrac{p - 2w}{2}$

**33.** $v = k + gt$

$v - k = gt$

$\dfrac{v - k}{g} = t$

$t = \dfrac{v - k}{g}$

**34.** $y = \dfrac{k}{z}$

$yz = k$

$z = \dfrac{k}{y}$

**35.** $V = lwh$

$\dfrac{V}{lw} = h$

$h = \dfrac{V}{lw}$

**36.** $W = I^2 R$

$\dfrac{W}{I^2} = R$

$R = \dfrac{W}{I^2}$

**37.** $180 = A + B + C$

$180 - A - C = B$

$B = 180 - A - C$

**38.** $S = \dfrac{a}{1 - r}$

$S(1 - r) = a$

$S - Sr = a$

$-Sr = a - S$

$r = \dfrac{a - S}{-S}$

or $r = \dfrac{S - a}{S}$

**39.** $A = \dfrac{h}{2}(b + c)$

$2A = h(b + c)$

$2A = hb + hc$

$2A - hb = hc$

$\dfrac{2A - hb}{h} = c$

$c = \dfrac{2A - hb}{h}$

**40.** $S = 2r(r + h)$

$S = 2r^2 + 2rh$

$S - 2r^2 = 2rh$

$\dfrac{S - 2r^2}{2r} = h$

$h = \dfrac{S - 2r^2}{2r}$

**Solution Key 1.4**

**41.**
$$S = 3\pi d + 5\pi D$$
$$S - 5\pi D = 3\pi d$$
$$\frac{S - 5\pi D}{3\pi} = d$$
$$d = \frac{S - 5\pi D}{3\pi}$$

**42.**
$$A = 2\pi rh + 2\pi r^2$$
$$A - 2\pi r^2 = 2\pi rh$$
$$\frac{A - 2\pi r^2}{2\pi r} = h$$
$$h = \frac{A - 2\pi r^2}{2\pi r}$$

## [1.4]

**1.** $\qquad 3x < 6$
Divide by 3. $\quad x < 2$
Solution set: $\{x \mid x < 2\}$

**2.** $\qquad 2x > 8$
Divide by 2. $\quad x > 4$
Solution set: $\{x \mid x > 4\}$

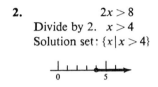

**3.** $\qquad x + 7 \geq 8$
Subtract 7. $\quad x \geq 1$
Solution set: $\{x \mid x \geq 1\}$

**4.** $\qquad x - 5 \leq 7$
Add 5. $\quad x \leq 12$
Solution set: $\{x \mid x \leq 12\}$

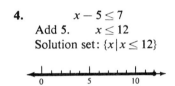

**5.** $\qquad -3x > 12$
Divide by $-3$. $\quad x < -4$
(Reverse sense of inequality)
Solution set: $\{x \mid x < -4\}$

**6.** $\qquad -4x < 20$
Divide by $-4$. $\quad x > -5$
(Reverse sense of inequality)
Solution set: $\{x \mid x > -5\}$

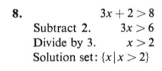

**7.** $\qquad 2x - 3 < 4$
Add 3. $\qquad 2x < 7$
Divide by 2. $\quad x < \frac{7}{2}$
Solution set: $\left\{x \mid x < \frac{7}{2}\right\}$

**8.** $\qquad 3x + 2 > 8$
Subtract 2. $\quad 3x > 6$
Divide by 3. $\quad x > 2$
Solution set: $\{x \mid x > 2\}$

**9.** $\qquad 3 - 2x > 12$
Subtract 3. $\quad -2x > 9$
Divide by $-2$. $\quad x < -\frac{9}{2}$
Solution set: $\left\{x \mid x < -\frac{9}{2}\right\}$

**10.** $\qquad 4 - 3x < 13$
Subtract 4. $\quad -3x < 9$
Divide by $-3$. $\quad x > -3$
Solution set: $\{x \mid x > -3\}$

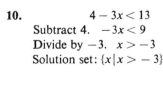

**11.**
$$2 - x < -3$$
Subtract 2.   $-x < -5$
Multiply by $-1$.   $x > 5$
Solution set: $\{x \mid x > 5\}$

**12.**
$$3 - x > -4$$
Subtract 3.   $-x > -7$
Multiply by $-1$.   $x < 7$
Solution set: $\{x \mid x < 7\}$

**13.**
$$3x - 2 \geq 1 + 2x$$
Add $-2x + 2$.   $x \geq 3$
Solution set: $\{x \mid x \geq 3\}$

**14.**
$$2x + 3 \leq x - 1$$
Add $-x - 3$.   $x \leq -4$
Solution set: $\{x \mid x \leq -4\}$

**15.**
$$\frac{2x - 6}{3} > 0$$
Multiply by 3.   $2x - 6 > 0$
Add 6.   $2x > 6$
Divide by 2.   $x > 3$
Solution set: $\{x \mid x > 3\}$

**16.**
$$\frac{2x - 3}{2} \leq 0$$
Multiply by 2.   $2x - 3 \leq 0$
Add 3.   $2x \leq 3$
Divide by 2.   $x \leq \frac{3}{2}$
Solution set: $\left\{x \mid x \leq \frac{3}{2}\right\}$

**17.**
$$\frac{5x - 7x}{3} \geq 0$$
$5x - 7x \geq 0$
$-2x \geq 0$
$x \leq 0$
Solution set: $\{x \mid x \leq 0\}$

**18.**
$$\frac{x - 3x}{5} \leq 6$$
$x - 3x \leq 30$
$-2x \leq 30$
$x \geq -15$
Solution set: $\{x \mid x \geq -15\}$

**19.**
$$\frac{R + 2}{3} < \frac{R - 2}{4}$$
Multiply by the least common denominator, 12.
$4(R + 2) < 3(R - 2)$
$4R + 8 < 3R - 6$
$R < -14$
Solution set: $\{R \mid R < -14\}$

**20.**
$$\frac{2R}{5} > R - \frac{1}{2}$$
Multiply by the least common denominator, 10.
$4R > 10R - 5$
$-6R > -5$
$R < \frac{5}{6}$
Solution set: $\left\{R \mid R < \frac{5}{6}\right\}$

**Solution Key 1.5**

**21.** $\dfrac{T-1}{3} \geq 2 + \dfrac{T}{2}$

Multiply by the least common denominator, 6.
$2(T-1) \geq 12 + 3T$
$2T - 2 \geq 12 + 3T$
$-14 \geq T$
$T \leq -14$
Solution set: $\{T | T \leq -14\}$

**22.** $\dfrac{2-T}{5} \leq -\dfrac{8+T}{3}$

Multiply by the least common denominator, 15.
$3(2-T) \leq -5(8+T)$
$6 - 3T \leq -40 - 5T$
$46 \leq -2T$
$-23 \geq T$
$T \leq -23$
Solution set: $\{T | T \leq -23\}$

**23.** $\dfrac{P-2P}{4} < \dfrac{2P}{3} + \dfrac{1}{2}$

Multiply by 12.
$3(P - 2P) < 8P + 6$
$3(-P) < 8P + 6$
$-3P < 8P + 6$
$-6 < 11P$
$\dfrac{-6}{11} < P$
$P > -\dfrac{6}{11}$
Solution set: $\left\{P \Big| P > -\dfrac{6}{11}\right\}$

**24.** $\dfrac{P}{2} - \dfrac{2P}{3} > \dfrac{P}{4}$

Multiply by 12.
$6P - 8P > 3P$
$-2P > 3P$
$0 > 5P$
$0 > P$
$P < 0$
Solution set: $\{P | P < 0\}$

## [1.5]*

**1.** Let  $n$ = first number
$3n - 2$ = second number
(a) $n + (3n - 2) = 30$
(b) $4n - 2 = 30$
$4n = 32$
$n = 8$
$3n - 2 = 22$
The numbers are 8 and 22.

**2.** Let $n$ = the number
(a) $\dfrac{3}{5}n = n - 8$
(b) $3n = 5n - 40$
$-2n = -40$
$n = 20$
The number is 20.

**3.** Let  $l$ = length of first piece of rope
$6l - 2$ = length of second piece of rope
(a) $l + (6l - 2) = 40$
(b) $7l - 2 = 40$
$7l = 42$
$l = 6$
$6l - 2 = 34$
The pieces of rope are 6 and 34 feet long.

**4.** Let  $w$ = number of votes for winner
$7179 - w$ = number of votes for loser
(a) $(w - 6) + 1 = (7179 - w) + 6$
(b) $w - 5 = 7185 - w$
$2w = 7190$
$w = 3595$
$7179 - w = 3584$
There were 3595 votes cast for the winner and 3584 votes for the loser.

* The methods shown for solving word problems here and in the following sections may not be the only ones possible. Your methods may be different and still be correct.

## Solution Key 1.5

**5.** Let $t$ = number of minutes each song lasts. With five songs there are four intervals of 15 seconds each between two songs, a total of 1 minute.
(a) $5t + 1 = 26$
(b) $5t = 25$
$t = 5$
Each song lasts 5 minutes.

**6.** Let $t$ = number of minutes for commercials
$30 - t$ = number of minutes for entertainment
(a) $30 - t = 3t + 2$
(b) $-4t = -28$
$t = 7$
7 minutes are devoted to commercials.

**7.** Let $x$ = first integer
$x + 1$ = second integer
$x + 2$ = third integer
(a) $x + (x + 1) + (x + 2) = 42$
(b) $3x + 3 = 42$
$3x = 39$
$x = 13$
$x + 1 = 14$
$x + 2 = 15$
The integers are 13, 14, and 15.

**8.** Let $x$ = lesser integer
$x + 1$ = greater integer
(a) $x + (x + 1) = 3x - 8$
(b) $2x + 1 = 3x - 8$
$9 = x$
$10 = x + 1$
The integers are 9 and 10.

**9.** Let $x$ = first integer
$x + 1$ = next integer
(a) $\frac{1}{3}x + \frac{1}{2}(x + 1) = 33$
(b) $2x + 3(x + 1) = 6(33)$
$5x + 3 = 198$
$5x = 195$
$x = 39$
$x + 1 = 40$
The integers are 39 and 40.

**10.** Let $x$ = first integer
$x + 1$ = next integer
(a) $\frac{1}{2}x + \frac{1}{5}(x + 1) = 17$
(b) $5x + 2(x + 1) = 10(17)$
$7x + 2 = 170$
$7x = 168$
$x = 24$
$x + 1 = 25$
The integers are 24 and 25.

**11.** Let $x$ = first integer
$x + 2$ = second integer
$x + 4$ = third integer
(a) $x + (x + 2) + (x + 4) = 63$
(b) $3x + 6 = 63$
$3x = 57$
$x = 19$
$x + 2 = 21$
$x + 4 = 23$
The integers are 19, 21, and 23.

**12.** Let $x$ = lesser integer
$x + 2$ = greater integer
(a) $-4x = x + (x + 2) + 46$
(b) $-4x = 2x + 48$
$-6x = 48$
$x = -8$
$x + 2 = -6$
The integers are $-8$ and $-6$.

**Solution Key 1.5**

**13.** Let $x =$ sister's age now
$3x =$ boy's age now
$x + 4 =$ sister's age in 4 years
$3x + 4 =$ boy's age in 4 years
(a) $3x + 4 = 2(x + 4)$
(b) $3x + 4 = 2x + 8$
$x = 4$
$3x = 12$
The boy is now 12 years old and his sister, 4.

**14.** Let $x =$ daughter's age now
$5x =$ woman's age now
$x + 9 =$ daughter's age in 9 years
$5x + 9 =$ woman's age in 9 years
(a) $5x + 9 = 3(x + 9)$
(b) $5x + 9 = 3x + 27$
$2x = 18$
$x = 9$
$5x = 45$
The woman is now 45 years old and her daughter, 9.

**15.** Let $x =$ sister's age now
$x + 4 =$ boy's age now
$x - 10 =$ sister's age 10 years ago
$x - 6 =$ boy's age 10 years ago
(a) $x - 6 = 2(x - 10)$
(b) $x - 6 = 2x - 20$
$14 = x$
$18 = x + 4$
The boy is now 18 years old and his sister, 14.

**16.** Let $x =$ age of son now
$x + 25 =$ age of father now
$x + 10 =$ age of son in 10 years
$x + 35 =$ age of father in 10 years
(a) $x + 35 = 2(x + 10)$
(b) $x + 35 = 2x + 20$
$15 = x$
$40 = x + 25$
The father is now 40 years old and his son, 15.

**17.** Let $x =$ age of son now
$60 - x =$ age of father now
$x + 4 =$ son's age in 4 years
$64 - x =$ father's age in 4 years
(a) $64 - x = 3(x + 4)$
(b) $64 - x = 3x + 12$
$52 = 4x$
$13 = x$
$47 = 60 - x$
The father is now 47 years old and his son, 13.

**18.** Let $x =$ Joe's age now
$45 - x =$ Pete's age now
$x + 5 =$ Joe's age in 5 years
$50 - x =$ Pete's age in 5 years
(a) $x + 5 = 4(50 - x)$
(b) $x + 5 = 200 - 4x$
$5x = 195$
$x = 39$
$45 - x = 6$
Joe's age now is 39 and Pete's is 6.

**19.** Let $x =$ number of nickels
$x + 3 =$ number of dimes
$5x =$ value of $x$ nickels in cents
$10(x + 3) =$ value of $x + 3$ dimes in cents
Value of nickels + value of dimes = total value
(a) $5x + 10(x + 3) = 180$
(b) $15x + 30 = 180$
$15x = 150$
$x = 10$
$x + 3 = 13$
There are 10 nickels and 13 dimes.

**20.** Let $x$ = number of quarters
$x + 7$ = number of dimes
$25x$ = value of $x$ quarters in cents
$10(x + 7)$ = value of $x + 7$ dimes in cents
Value of quarters + value of dimes = total value
(a) $25x + 10(x + 7) = 280$
(b) $25x + 10x + 70 = 280$
$35x = 210$
$x = 6$
$x + 7 = 13$
There are 6 quarters and 13 dimes.

**21.** Let $x$ = number of nickels
$16 - x$ = number of quarters
$5x$ = value of nickels in cents
$25(16 - x)$ = value of quarters in cents
Value of nickels + value of quarters = total value
(a) $5x + 25(16 - x) = 220$
(b) $5x + 400 - 25x = 220$
$-20x = -180$
$x = 9$
$16 - x = 7$
There are 9 nickels and 7 quarters.

**22.** Let $x$ = number of $10 bills
$x + 10$ = number of $5 bills
$94 - x - (x + 10)$ = number of $1 bills
$10x$ = value of $10 bills in dollars
$5(x + 10)$ = value of $5 bills in dollars
$[94 - x - (x + 10)]$ = value of $1 bills in dollars
Value of $10 bills + value of $5 bills + value of $1 bills = total value
(a) $10x + 5(x + 10) + [94 - x - (x + 10)] = 446$
(b) $10x + 5x + 50 + 94 - x - x - 10 = 446$
$13x + 134 = 446$
$x = 24$
$x + 10 = 34$
$94 - x - (x + 10) = 36$
There are 24 ten-, 34 five-, and 36 one-dollar bills

**23.** Let $x$ = number of bars bought
$x - 1$ = number of bars sold
$10(x - 1)$ = value of bars sold in cents
$\left[\text{Note: Cost for 3 bars is 20¢ or } \dfrac{20}{3}\text{¢ each.}\right]$
Value of bars sold − cost of bars bought = profit
(a) $10(x - 1) - \dfrac{20}{3}x = 200$
$30(x - 1) - 20x = 600$
(b) $30x - 30 - 20x = 600$
$10x = 630$
$x = 63$
He bought 63 bars.

**Solution Key 1.5**

**24.** Let $x =$ number of children
$82 - x =$ number of adults
$85x =$ value of children's tickets in cents
$150(82 - x) =$ value of adults' tickets in cents
Value of children's tickets + value of adults' tickets = total value
(a) $85x + 150(82 - x) = 9310$
(b) $85x + 12{,}300 - 150x = 9310$
$-65x = -2990$
$x = 46$
$82 - x = 36$
46 children and 36 adults attended.

**25.** Let $A =$ amount invested at 6%
$2000 - A =$ amount invested at 8%
$0.06A =$ dollars of interest on 6% investment
$0.08(2000 - A) =$ dollars of interest on 8% investment
Interest on 6% amount + interest on 8% amount = total interest
(a) $0.06A + 0.08(2000 - A) = 132$
(b) $6A + 8(2000 - A) = 13{,}200$
$6A + 16{,}000 - 8A = 13{,}200$
$-2A = -2800$
$A = 1400$
$2000 - A = 600$
$1400 at 6% and $600 at 8%.

**26.** Let $A =$ amount invested at 5%
$2200 - A =$ amount invested at 6%
$0.05A =$ dollars of interest on 5% investment
$0.06(2200 - A) =$ dollars of interest on 6% investment
Interest on 5% investment = interest on 6% investment
(a) $0.05A = 0.06(2200 - A)$
(b) $5A = 6(2200 - A)$
$5A = 13{,}200 - 6A$
$11A = 13{,}200$
$A = 1200 \qquad 0.05A = 60$ (interest on 5% investment)
$2200 - A = 1000 \quad 0.06(2200 - A) = 60$ (interest on 6% investment)
The yearly interest on both is $120.

**27.** Let $A =$ amount invested at each rate
$0.05A =$ dollars of interest on amount invested at 5%
$0.07A =$ dollars of interest on amount invested at 7%
Interest from 5% investment + interest from 7% investment = total interest
(a) $0.05A + 0.07A = 144$
(b) $5A + 7A = 14{,}400$
$12A = 14{,}400$
$A = 1200$
He has $1200 invested at each rate.

**28.** Let $\quad A =$ amount invested in stocks paying 5%
$\quad\quad\quad\quad 3A =$ amount invested in bonds paying 8%
$\quad\quad\quad\quad 0.05A =$ dollars of interest on stocks
$\quad\quad\quad\quad 0.08(3A) =$ dollars of interest on bonds

Interest from stock investment + interest from bond investment = total interest

(a) $\quad 0.05A + 0.08(3A) = 1740$
(b) $\quad\quad\quad 5A + 8(3A) = 174{,}000$
$\quad\quad\quad\quad\quad 29A = 174{,}000$
$\quad\quad\quad\quad\quad\quad A = 6{,}000$
$\quad\quad\quad\quad\quad 3A = 18{,}000$

He has $6000 in stocks and $18,000 in bonds.

**29.** Let $\quad\quad A =$ amount invested at 8%
$\quad\quad A + 1000 =$ amount invested at 5%
$\quad\quad\quad 0.08A =$ dollars of interest on 8% investment
$\quad 0.05(A + 1000) =$ dollars of interest on 5% investment

Interest from 8% investment + interest from 5% investment = total interest

(a) $\quad 0.08A + 0.05(A + 1000) = 739$
(b) $\quad\quad\quad 8A + 5(A + 1000) = 73{,}900$
$\quad\quad\quad 8A + 5A + 5000 = 73{,}900$
$\quad\quad\quad\quad\quad\quad 13A = 68{,}900$
$\quad\quad\quad\quad\quad\quad\quad A = 5300$
$\quad\quad\quad\quad\quad A + 1000 = 6300$

He has $5300 at 8% and $6300 at 5%.

**30.** Let $\quad\quad A =$ amount invested at 8%
$\quad\quad A + 500 =$ amount invested at 6%
$\quad\quad\quad 0.08A =$ dollars of interest on 8% investment
$\quad 0.06(A + 500) =$ dollars of interest on 6% investment

Interest from 8% investment + interest from 6% investment = total interest

(a) $\quad 0.08A + 0.06(A + 500) = 114$
(b) $\quad\quad 8A + 6(A + 500) = 11{,}400$
$\quad\quad 8A + 6A + 3000 = 11{,}400$
$\quad\quad\quad\quad\quad 14A = 8400$
$\quad\quad\quad\quad\quad\quad A = 600$
$\quad\quad\quad\quad A + 500 = 1100$

He has $600 at 8% and $1100 at 6%.

**31.** Let $\quad\quad r =$ rate of the automobile
$\quad\quad r + 120 =$ rate of airplane
Time of auto = time of plane
$$\frac{\text{Distance of auto}}{\text{Rate of auto}} = \frac{\text{distance of plane}}{\text{rate of plane}} \quad \left[\text{Time} = \frac{\text{distance}}{\text{rate}}\right]$$

(a) $\quad\dfrac{420}{r} = \dfrac{1260}{r + 120}$

(b) $\quad r(r+120)\left(\dfrac{420}{r}\right) = r(r+120)\left(\dfrac{1260}{r+120}\right)$
$\quad\quad (r + 120)420 = r(1260)$
$\quad\quad 420r + 50{,}400 = 1260r$
$\quad\quad\quad\quad\quad 60 = r$
$\quad\quad\quad\quad 180 = r + 120$

Automobile, 60 mph; airplane, 180 mph.

Solution Key 1.5

**32.** Let $r$ = rate of slower car.
$2r$ = rate of faster car.
Distance of faster car = distance of slower car + 96   (Distance = rate × time)
(Rate of faster car) × (time of faster car) = (rate of slower car) × (time of slower car) + 96
(a)  $(2r)(3) = (r)(3) + 96$
(b)   $6r = 3r + 96$
      $3r = 96$
      $r = 32$
      $2r = 64$
Slower car, 32 mph; faster car, 64 mph.

**33.** Let  $t$ = time freight train travels
      $t - 3$ = time passenger train travels
Distance of freight = distance of passenger   (Distance = rate × time)
(Rate of freight) × (time of freight) = (rate of passenger) × (time of passenger)
(a)   $40t = 80(t - 3)$
(b)   $40t = 80t - 240$
      $-40t = -240$
      $t = 6$
Distance = $40(6) = 240$
The point is 240 miles from town A.

**34.** Let $t$ = time each plane travels   (Distance = rate × time)
Total distance = (rate of first plane) × $t$ + (rate of second plane) × $t$
(a)   $2880 = 520t + 440t$
(b)   $2880 = 960t$
      $3 = t$
In 3 hours they will be 2880 miles apart.

**35.** Let  $t$ = time slow car travels
      $t - 2$ = time fast car travels
Distance of slow car = distance of fast car   (Distance = rate × time)
(Rate of slow car) × (time of slow car) = (rate of fast car) × (time of fast car)
(a)   $42t = 54(t - 2)$
(b)   $42t = 54t - 108$
      $-12t = -108$
      $t = 9$
Distance = $42(9) = 378$
The point is 378 miles from town.

**36.** Let  $t$ = time slow boat travels
      $t - 1$ = time fast boat travels
Distance of slow boat = distance of fast boat   (Distance = rate × time)
(Rate of slow boat) × (time of slow boat) = (rate of fast boat) × (time of fast boat)
(a)   $36t = 45(t - 1)$
(b)   $36t = 45t - 45$
      $-9t = -45$
      $t = 5$
Distance = $36(5) = 180$
The point is 180 nautical miles from the harbor.

**Solution Key 1.5**

**37.** Let $x$ = grade (in %) on the fifth test
  (a) $80 \leq \dfrac{78 + 64 + 88 + 76 + x}{5} < 90$
  (b) $400 \leq 306 + x < 450$
  $94 \leq x < 144$
  Any grade equal to or greater than 94% would give the student a B. (He cannot exceed 100%.)

**38.** $\quad -9 < C < 27$
  (a) $-9 < \dfrac{5}{9}(F - 32) < 27$
  (b) $-81 < 5(F - 32) < 243$
  $-81 < 5F - 160 < 243$
  $79 < 5F < 403$
  $\dfrac{79}{5} < F < \dfrac{403}{5}$
  The temperature must be between $15\tfrac{4}{5}°$ and $80\tfrac{3}{5}°$ Fahrenheit.

**39.** Let $\qquad A$ = amount invested at 7%
  $10{,}000 - A$ = amount invested at 5%
  $0.07A$ = dollars of interest on 7% investment
  $0.05(10{,}000 - A)$ = dollars of interest on 5% investment
  Interest on 7% investment + interest on 5% investment $\geq 616$
  (a) $0.07A + 0.05(10{,}000 - A) \geq 616$
  (b) $\qquad 7A + 5(10{,}000 - A) \geq 61{,}600$
  $\qquad 7A + 50{,}000 - 5A \geq 61{,}600$
  $\qquad\qquad\qquad\quad 2A \geq 11{,}600$
  $\qquad\qquad\qquad\quad\; A \geq 5800$
  He must invest $5800 or more at 7% in order to earn at least $616 annual interest.

**40.** Let $d$ = distance he sails upstream. At 4 mph, it will take him $d/4$ hours upstream, and at 6 mph, it will take him $d/6$ hours downstream. Now from 8 A.M. to 6 P.M. is an interval of 10 hours. Hence,
  (a) $\quad \dfrac{d}{4} + \dfrac{d}{6} \leq 10$
  (b) $\; 12\left(\dfrac{d}{4} + \dfrac{d}{6}\right) \leq 12(10)$
  $\qquad 3d + 2d \leq 120$
  $\qquad\qquad 5d \leq 120$
  $\qquad\qquad\; d \leq 24$
  Thus he can sail, at most, 24 miles upstream.

# Unit 1 Review

**1.** $\dfrac{2x+1}{5} - x = 4$

Does $\dfrac{2(-3)+1}{5} - (-3) = 4$?

Does $-\dfrac{5}{5} + 3 = 4$?

Does $-1 + 3 = 4$?

No.

**2.** $4x - 2 = 6x + 4$

Does $4(-3) - 2 = 6(-3) + 4$?

Does $-12 - 2 = -18 + 4$?

Does $\phantom{-12 - 2 =} -14 = -14$?

Yes.

**3.** $2(x - 3) = 12$
$2x - 6 = 12$
$2x = 18$
$x = 9$

Solution set: $\{9\}$

**4.** $3[x - 2(x + 1)] = 4 + x$
$3[x - 2x - 2] = 4 + x$
$3[-x - 2] = 4 + x$
$-3x - 6 = 4 + x$
$-4x = 10$
$x = -\dfrac{10}{4}$
$x = -\dfrac{5}{2}$

Solution set: $\left\{-\dfrac{5}{2}\right\}$

**5.** $\dfrac{x}{4} - 2 = \dfrac{x+1}{3}$

$12\left(\dfrac{x}{4} - 2\right) = 12\left(\dfrac{x+1}{3}\right)$

$3x - 24 = 4(x + 1)$
$3x - 24 = 4x + 4$
$-28 = x$

Solution set: $\{-28\}$

**6.** $\dfrac{2}{x-1} + \dfrac{1}{3(x-1)} = \dfrac{7}{2}$

$6(x-1)\left(\dfrac{2}{x-1} + \dfrac{1}{3(x-1)}\right) = 6(x-1)\dfrac{7}{2}$

$12 + 2 = 21(x-1)$
$14 = 21x - 21$
$35 = 21x$
$\dfrac{35}{21} = x$
$\dfrac{5}{3} = x$

Solution set: $\left\{\dfrac{5}{3}\right\}$

**7.** $\dfrac{2t+6}{3} = \dfrac{1}{2} + t$

$6\left(\dfrac{2t+6}{3}\right) = 6\left(\dfrac{1}{2} + t\right)$

$2(2t + 6) = 3 + 6t$
$4t + 12 = 3 + 6t$
$9 = 2t$
$\dfrac{9}{2} = t$

Solution set: $\left\{\dfrac{9}{2}\right\}$

**8.** $\dfrac{1}{r} + \dfrac{1}{2r} = \dfrac{3}{2}$

$2r\left(\dfrac{1}{r} + \dfrac{1}{2r}\right) = 2r\left(\dfrac{3}{2}\right)$

$2 + 1 = 3r$
$3 = 3r$
$1 = r$

Solution set: $\{1\}$

## Solution Key Unit 1 Review

**9.** $I = prt$
$\dfrac{I}{pt} = r$
$r = \dfrac{I}{pt}$

**10.** $r = k + gt$
$r - k = gt$
$\dfrac{r-k}{t} = g$
$g = \dfrac{r-k}{t}$

**11.** $S = 3\pi d + 5\pi D$
$S - 3\pi d = 5\pi D$
$\dfrac{S - 3\pi d}{5\pi} = D$
$D = \dfrac{S - 3\pi d}{5\pi}$

**12.** $\dfrac{1}{r} = \dfrac{1}{r_1} + \dfrac{1}{r_2}$
$rr_1r_2\left(\dfrac{1}{r}\right) = rr_1r_2\left(\dfrac{1}{r_1} + \dfrac{1}{r_2}\right)$
$r_1r_2 = rr_2 + rr_1$
$r_1r_2 = r(r_2 + r_1)$
$\dfrac{r_1r_2}{r_2 + r_1} = r$
$r = \dfrac{r_1r_2}{r_2 + r_1}$

**13.** $\dfrac{2x}{3} \geq 4$
$2x \geq 12$
$x \geq 6$
Solution set: $\{x \mid x \geq 6\}$

**14.** $-3x + 1 < 7$
$-3x < 6$
$x > -2$
Solution set: $\{x \mid x > -2\}$

**15.** $\dfrac{3x - 2}{2} \leq 5$
$2\left(\dfrac{3x-2}{2}\right) \leq 2(5)$
$3x - 2 \leq 10$
$3x \leq 12$
$x \leq 4$
Solution set: $\{x \mid x \leq 4\}$

**16.** $\dfrac{2}{5} + x > \dfrac{x}{2}$
$10\left(\dfrac{2}{5} + x\right) > 10\left(\dfrac{x}{2}\right)$
$4 + 10x > 5x$
$5x > -4$
$x > -\dfrac{4}{5}$
Solution set: $\left\{x \mid x > -\dfrac{4}{5}\right\}$

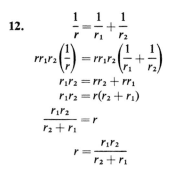

**17.** Let $r$ = rate of investment of the remainder of $5000.
Interest on $5000 at 4% + interest on $10,000 at 3% + interest on remainder = 800
$0.04(5000) + 0.03(10,000) + r(5000) = 800$
$200 + 300 + 5000r = 800$
$5000r = 300$
$r = \dfrac{300}{5000}$
$r = 0.06$
He should invest the remainder at 6%.

Solution Key 2.1

**18.** Let $\quad x =$ number of dimes
$x + 2 =$ number of nickels
$10x =$ value of dimes in cents
$5(x + 2) =$ value of nickels in cents

Value of dimes + value of nickels = total value

$$10x + 5(x + 2) = 160$$
$$10x + 5x + 10 = 160$$
$$15x = 150$$
$$x = 10$$
$$x + 2 = 12$$

He has 10 dimes and 12 nickels.

**19.** Let $\quad r =$ rate of slower train
$r + 6 =$ rate of faster train

Time of slow train = time of fast train

$$\frac{\text{Distance of slow train}}{\text{Rate of slow train}} = \frac{\text{distance of fast train}}{\text{rate of fast train}} \qquad \left[\text{Time} = \frac{\text{distance}}{\text{rate}}\right]$$

$$\frac{210}{r} = \frac{280}{r + 6}$$
$$r(r + 6)\left(\frac{210}{r}\right) = r(r + 6)\left(\frac{280}{r + 6}\right)$$
$$210(r + 6) = 280r$$
$$210r + 1260 = 280r$$
$$1260 = 70r$$
$$18 = r$$
$$24 = r + 6$$

First freight, 24 mph; second freight, 18 mph

**20.** Let $\quad A =$ amount invested at 6%
$0.06A =$ dollars of interest on 6% investment
$5000 + A =$ total amount invested

Interest on $5000 at 4% + interest on 6% investment $\geq$ interest at 5% on total investment

$$0.04(5000) + 0.06A \geq 0.05(5000 + A)$$
$$4(5000) + 6A \geq 5(5000 + A)$$
$$20{,}000 + 6A \geq 25{,}000 + 5A$$
$$A \geq 5000$$

He must invest at least $5000 in order to earn at least 5% on the total investment.

## [2.1]

**1.** $(x + 2)(x - 5)$
$x + 2 = 0; \quad x - 5 = 0$
$\{-2, 5\}$

**2.** $(x + 3)(x - 4)$
$x + 3 = 0; \quad x - 4 = 0$
$\{-3, 4\}$

3.  $(2x+5)(x-2)$
    $2x+5=0; \quad x-2=0$
    $\left\{-\dfrac{5}{2}, 2\right\}$

4.  $(x+1)(3x-1)$
    $x+1=0; \quad 3x-1=0$
    $\left\{-1, \dfrac{1}{3}\right\}$

5.  $x(2x+1)$
    $x=0; \quad 2x+1=0$
    $\left\{0, \dfrac{-1}{2}\right\}$

6.  $x(3x-7)$
    $x=0; \quad 3x-7=0$
    $\left\{0, \dfrac{7}{3}\right\}$

7.  $4(x-6)(2x+3)$
    $x-6=0; \quad 2x+3=0$
    $\left\{6, \dfrac{-3}{2}\right\}$

8.  $5(2x-7)(x+1)$
    $2x-7=0; \quad x+1=0$
    $\left\{\dfrac{7}{2}, -1\right\}$

9.  $(x-2)(2x+1)=0$
    $x-2=0; \quad 2x+1=0$
    $\left\{2, \dfrac{-1}{2}\right\}$

10. $(x+5)(3x-1)=0$
    $x+5=0; \quad 3x-1=0$
    $\left\{-5, \dfrac{1}{3}\right\}$

11. $4(2x-5)(3x+2)=0$
    $2x-5=0; \quad 3x+2=0$
    $\left\{\dfrac{5}{2}, \dfrac{-2}{3}\right\}$

12. $2(3x-4)(2x+5)=0$
    $3x-4=0; \quad 2x+5=0$
    $\left\{\dfrac{4}{3}, \dfrac{-5}{2}\right\}$

13. $4x(2x-9)=0$
    $x=0; \quad 2x-9=0$
    $\left\{0, \dfrac{9}{2}\right\}$

14. $3x(5x+1)=0$
    $x=0; \quad 5x+1=0$
    $\left\{0, \dfrac{-1}{5}\right\}$

15. $x^2-3x=0$
    $x(x-3)=0$
    $x=0; \quad x-3=0$
    $\{0,3\}$

16. $x^2+5x=0$
    $x(x+5)=0$
    $x=0; \quad x+5=0$
    $\{0,-5\}$

17. $\quad 2x^2=6x$
    $2x^2-6x=0$
    $2x(x-3)=0$
    $x=0; \quad x-3=0$
    $\{0,3\}$

18. $\quad 3x^2=3x$
    $3x^2-3x=0$
    $3x(x-1)=0$
    $x=0; \quad x-1=0$
    $\{0,1\}$

19. $\quad x^2-9=0$
    $(x+3)(x-3)=0$
    $x+3=0; \quad x-3=0$
    $\{-3,3\}$

20. $\quad x^2-4=0$
    $(x+2)(x-2)=0$
    $x+2=0; \quad x-2=0$
    $\{-2,2\}$

21. $\quad 2x^2-18=0$
    $\quad 2(x^2-9)=0$
    $2(x+3)(x-3)=0$
    $x+3=0; \quad x-3=0$
    $\{-3,3\}$

22. $\quad 3x^2-3=0$
    $\quad 3(x^2-1)=0$
    $3(x+1)(x-1)=0$
    $x+1=0; \quad x-1=0$
    $\{-1,1\}$

**23.**
$$9x^2 = 4$$
$$9x^2 - 4 = 0$$
$$(3x+2)(3x-2) = 0$$
$$3x+2=0; \quad 3x-2=0$$
$$\left\{\frac{-2}{3}, \frac{2}{3}\right\}$$

**24.**
$$4x^2 = 1$$
$$4x^2 - 1 = 0$$
$$(2x+1)(2x-1) = 0$$
$$2x+1=0; \quad 2x-1=0$$
$$\left\{\frac{-1}{2}, \frac{1}{2}\right\}$$

**25.**
$$\frac{4}{9}x^2 - 1 = 0$$
$$9\left(\frac{4}{9}x^2 - 1\right) = (9)0$$
$$4x^2 - 9 = 0$$
$$(2x+3)(2x-3) = 0$$
$$2x+3=0; \quad 2x-3=0$$
$$\left\{\frac{-3}{2}, \frac{3}{2}\right\}$$

**26.**
$$\frac{9x^2}{25} - 1 = 0$$
$$25\left(\frac{9x^2}{25} - 1\right) = (25)0$$
$$9x^2 - 25 = 0$$
$$(3x+5)(3x-5) = 0$$
$$3x+5=0; \quad 3x-5=0$$
$$\left\{\frac{-5}{3}, \frac{5}{3}\right\}$$

**27.**
$$x^2 - 5x + 4 = 0$$
$$(x-1)(x-4) = 0$$
$$x-1=0; \quad x-4=0$$
$$\{1,4\}$$

**28.**
$$x^2 + 5x + 6 = 0$$
$$(x+2)(x+3) = 0$$
$$x+2=0; \quad x+3=0$$
$$\{-2,-3\}$$

**29.**
$$x^2 - 5x - 14 = 0$$
$$(x+2)(x-7) = 0$$
$$x+2=0; \quad x-7=0$$
$$\{-2,7\}$$

**30.**
$$x^2 - x - 42 = 0$$
$$(x+6)(x-7) = 0$$
$$x+6=0; \quad x-7=0$$
$$\{-6,7\}$$

**31.**
$$3x^2 - 6x = -3$$
$$3x^2 - 6x + 3 = 0$$
$$3(x^2 - 2x + 1) = 0$$
$$3(x-1)(x-1) = 0$$
$$x-1=0$$
$$\{1\}$$

**32.**
$$12x^2 = 8x + 15$$
$$12x^2 - 8x - 15 = 0$$
$$(2x-3)(6x+5) = 0$$
$$2x-3=0; \quad 6x+5=0$$
$$\left\{\frac{3}{2}, \frac{-5}{6}\right\}$$

**33.**
$$x(2x-3) = -1$$
$$2x^2 - 3x + 1 = 0$$
$$(2x-1)(x-1) = 0$$
$$2x-1=0; \quad x-1=0$$
$$\left\{\frac{1}{2}, 1\right\}$$

**34.**
$$2x(x-2) = x+3$$
$$2x^2 - 4x = x+3$$
$$2x^2 - 5x - 3 = 0$$
$$(2x+1)(x-3) = 0$$
$$2x+1=0; \quad x-3=0$$
$$\left\{\frac{-1}{2}, 3\right\}$$

**35.**
$$(x-2)(x+1) = 4$$
$$x^2 - x - 2 = 4$$
$$x^2 - x - 6 = 0$$
$$(x+2)(x-3) = 0$$
$$x+2=0; \quad x-3=0$$
$$\{-2,3\}$$

**36.**
$$x(3x+2) = (x+2)^2$$
$$3x^2 + 2x = x^2 + 4x + 4$$
$$2x^2 - 2x - 4 = 0$$
$$2(x^2 - x - 2) = 0$$
$$2(x+1)(x-2) = 0$$
$$x+1=0; \quad x-2=0$$
$$\{-1,2\}$$

**37.** $\dfrac{2x^2}{3} + \dfrac{x}{3} - 2 = 0$

$3\left(\dfrac{2x^2}{3} + \dfrac{x}{3} - 2\right) = (3)0$

$2x^2 + x - 6 = 0$

$(x+2)(2x-3) = 0$

$x+2 = 0; \quad 2x-3 = 0$

$\left\{-2, \dfrac{3}{2}\right\}$

**38.** $x - 1 = \dfrac{x^2}{4}$

$4(x-1) = (4)\dfrac{x^2}{4}$

$4x - 4 = x^2$

$x^2 - 4x + 4 = 0$

$(x-2)(x-2) = 0$

$x - 2 = 0$

$\{2\}$

**39.** $\dfrac{x^2}{6} + \dfrac{x}{3} = \dfrac{1}{2}$

$6\left(\dfrac{x^2}{6} + \dfrac{x}{3}\right) = 6\left(\dfrac{1}{2}\right)$

$x^2 + 2x = 3$

$x^2 + 2x - 3 = 0$

$(x+3)(x-1) = 0$

$x+3 = 0; \quad x-1 = 0$

$\{-3, 1\}$

**40.** $\dfrac{x}{4} - \dfrac{3}{4} = \dfrac{1}{x}$

$4x\left(\dfrac{x}{4} - \dfrac{3}{4}\right) = 4x\left(\dfrac{1}{x}\right)$

$x^2 - 3x = 4$

$x^2 - 3x - 4 = 0$

$(x+1)(x-4) = 0$

$x+1 = 0; \quad x-4 = 0$

$\{-1, 4\}$

**41.** $3 = \dfrac{10}{x^2} - \dfrac{7}{x}$

$x^2(3) = x^2\left(\dfrac{10}{x^2} - \dfrac{7}{x}\right)$

$3x^2 = 10 - 7x$

$3x^2 + 7x - 10 = 0$

$(3x+10)(x-1) = 0$

$3x+10 = 0; \quad x-1 = 0$

$\left\{\dfrac{-10}{3}, 1\right\}$

**42.** $\dfrac{4}{3x} + \dfrac{3}{3x+1} + 2 = 0$

$[3x(3x+1)]\dfrac{4}{3x} + [3x(3x+1)]\dfrac{3}{(3x+1)} + [3x(3x+1)]2 = [3x(3x+1)]0$

$4(3x+1) + (3x)3 + 6x(3x+1) = 0$

$18x^2 + 27x + 4 = 0$

$(3x+4)(6x+1) = 0$

$3x+4 = 0; \quad 6x+1 = 0$

$\left\{\dfrac{-4}{3}, \dfrac{-1}{6}\right\}$

**43.** $\dfrac{2}{x-3} - \dfrac{6}{x-8} = -1$

$[(x-3)(x-8)]\dfrac{2}{(x-3)} - [(x-3)(x-8)]\dfrac{6}{(x-8)} = [(x-3)(x-8)](-1)$

$2(x-8) - 6(x-3) = -(x^2 - 11x + 24)$

$x^2 - 15x + 26 = 0$

$(x-2)(x-13) = 0$

$x-2 = 0; \quad x-13 = 0$

$\{2, 13\}$

Solution Key 2.2

**44.**
$$\frac{x}{x-1} - \frac{x}{x+1} = \frac{4}{3}$$
$$[3(x-1)(x+1)]\frac{x}{(x-1)} - [3(x-1)(x+1)]\frac{x}{(x+1)} = [3(x-1)(x+1)]\frac{4}{3}$$
$$3x(x+1) - 3x(x-1) = 4(x-1)(x+1)$$
$$4x^2 - 6x - 4 = 0$$
$$2(2x^2 - 3x - 2) = 0$$
$$2(2x+1)(x-2) = 0$$
$$2x+1 = 0; \quad x-2 = 0$$
$$\left\{\frac{-1}{2}, 2\right\}$$

[2.2]

**1.** $x^2 = 9$
$x = 3; \quad x = -3$
$\{3, -3\}$

**2.** $x^2 = 16$
$x = 4; \quad x = -4$
$\{4, -4\}$

**3.** $9x^2 = 25$
$x^2 = \frac{25}{9}$
$x = \frac{5}{3}; \quad x = -\frac{5}{3}$
$\left\{\frac{5}{3}, -\frac{5}{3}\right\}$

**4.** $4x^2 = 9$
$x^2 = \frac{9}{4}$
$x = \frac{3}{2}; \quad x = -\frac{3}{2}$
$\left\{\frac{3}{2}, -\frac{3}{2}\right\}$

**5.** $2x^2 = 14$
$x^2 = 7$
$x = \sqrt{7}; \quad x = -\sqrt{7}$
$\{\sqrt{7}, -\sqrt{7}\}$

**6.** $3x^2 = 15$
$x^2 = 5$
$x = \sqrt{5}; \quad x = -\sqrt{5}$
$\{\sqrt{5}, -\sqrt{5}\}$

**7.** $4x^2 - 24 = 0$
$4x^2 = 24$
$x^2 = 6$
$x = \sqrt{6}; \quad x = -\sqrt{6}$
$\{\sqrt{6}, -\sqrt{6}\}$

**8.** $3x^2 - 9 = 0$
$3x^2 = 9$
$x^2 = 3$
$x = \sqrt{3}; \quad x = -\sqrt{3}$
$\{\sqrt{3}, -\sqrt{3}\}$

**9.** $\frac{2x^2}{3} = 4$
$2x^2 = 12$
$x^2 = 6$
$x = \sqrt{6}; \quad x = -\sqrt{6}$
$\{\sqrt{6}, -\sqrt{6}\}$

**10.** $\frac{3x^2}{5} = 3$
$3x^2 = 15$
$x^2 = 5$
$x = \sqrt{5}; \quad x = -\sqrt{5}$
$\{\sqrt{5}, -\sqrt{5}\}$

11. $\dfrac{5x^2}{2} - 5 = 0$
$\dfrac{5x^2}{2} = 5$
$5x^2 = 10$
$x^2 = 2$
$x = \sqrt{2}; \quad x = -\sqrt{2}$
$\{\sqrt{2}, -\sqrt{2}\}$

12. $\dfrac{7x^2}{3} - 14 = 0$
$\dfrac{7x^2}{3} = 14$
$7x^2 = 42$
$x^2 = 6$
$x = \sqrt{6}; \quad x = -\sqrt{6}$
$\{\sqrt{6}, -\sqrt{6}\}$

13. $(x - 2)^2 = 9$
$x - 2 = 3; \quad x - 2 = -3$
$x = 5; \qquad x = -1$
$\{5, -1\}$

14. $(x + 3)^2 = 4$
$x + 3 = 2; \quad x + 3 = -2$
$x = -1; \qquad x = -5$
$\{-1, -5\}$

15. $(2x - 1)^2 = 16$
$2x - 1 = 4; \quad 2x - 1 = -4$
$2x = 5; \qquad 2x = -3$
$x = \dfrac{5}{2}; \qquad x = -\dfrac{3}{2}$
$\left\{\dfrac{5}{2}, \dfrac{-3}{2}\right\}$

16. $(3x + 1)^2 = 25$
$3x + 1 = 5; \quad 3x + 1 = -5$
$3x = 4; \qquad 3x = -6$
$x = \dfrac{4}{3}; \qquad x = -2$
$\left\{\dfrac{4}{3}, -2\right\}$

17. $(x + 2)^2 = 3$
$x + 2 = \sqrt{3}; \qquad x + 2 = -\sqrt{3}$
$x = -2 + \sqrt{3}; \quad x = -2 - \sqrt{3}$
$\{-2 + \sqrt{3}, -2 - \sqrt{3}\}$

18. $(x - 5)^2 = 7$
$x - 5 = \sqrt{7}; \quad x - 5 = -\sqrt{7}$
$x = 5 + \sqrt{7}; \quad x = 5 - \sqrt{7}$
$\{5 + \sqrt{7}, 5 - \sqrt{7}\}$

19. $(x - 2)^2 = 12$
$x - 2 = \sqrt{12}; \quad x - 2 = -\sqrt{12}$
$x - 2 = 2\sqrt{3}; \quad -2 = -2\sqrt{3}$
$x = 2 + 2\sqrt{3}; \quad x = 2 - 2\sqrt{3}$
$\{2 + 2\sqrt{3}, 2 - 2\sqrt{3}\}$

20. $(x + 3)^2 = 18$
$x + 3 = \sqrt{18}; \qquad x + 3 = -\sqrt{18}$
$x + 3 = 3\sqrt{2}; \qquad x + 3 = -3\sqrt{2}$
$x = -3 + 3\sqrt{2}; \quad x = -3 - 3\sqrt{2}$
$\{-3 + 3\sqrt{2}, -3 - 3\sqrt{2}\}$

## [2.3]

1. (a) $\left(\dfrac{2}{2}\right)^2 = 1$    (b) $x^2 + 2x + 1 = (x + 1)^2$

2. (a) $\left(\dfrac{-4}{2}\right)^2 = 4$    (b) $x^2 - 4x + 4 = (x - 2)^2$

3. (a) $\left(\dfrac{-6}{2}\right)^2 = 9$    (b) $x^2 - 6x + 9 = (x - 3)^2$

4. (a) $\left(\dfrac{10}{2}\right)^2 = 25$    (b) $x^2 + 10x + 25 = (x + 5)^2$

5. (a) $\left(\dfrac{3}{2}\right)^2 = \dfrac{9}{4}$    (b) $x^2 + 3x + \dfrac{9}{4} = \left(x + \dfrac{3}{2}\right)^2$

Solution Key  2.3

6. (a) $\left(\dfrac{-5}{2}\right)^2 = \dfrac{25}{4}$  (b) $x^2 - 5x + \dfrac{25}{4} = \left(x - \dfrac{5}{2}\right)^2$

7. (a) $\left(\dfrac{-7}{2}\right)^2 = \dfrac{49}{4}$  (b) $x^2 - 7x + \dfrac{49}{4} = \left(x - \dfrac{7}{2}\right)^2$

8. (a) $\left(\dfrac{11}{2}\right)^2 = \dfrac{121}{4}$  (b) $x^2 + 11x + \dfrac{121}{4} = \left(x + \dfrac{11}{2}\right)^2$

9. (a) $\left(\dfrac{-1}{2}\right)^2 = \dfrac{1}{4}$  (b) $x^2 - x + \dfrac{1}{4} = \left(x - \dfrac{1}{2}\right)^2$

10. (a) $\left(\dfrac{15}{2}\right)^2 = \dfrac{225}{4}$  (b) $x^2 + 15x + \dfrac{225}{4} = \left(x + \dfrac{15}{2}\right)^2$

11. (a) $\left(\dfrac{\frac{1}{2}}{2}\right)^2 = \dfrac{1}{16}$  (b) $x^2 + \dfrac{1}{2}x + \dfrac{1}{16} = \left(x + \dfrac{1}{4}\right)^2$

12. (a) $\left(\dfrac{-\frac{3}{4}}{2}\right)^2 = \dfrac{9}{64}$  (b) $x^2 - \dfrac{3}{4}x + \dfrac{9}{64} = \left(x - \dfrac{3}{8}\right)^2$

13. $x^2 + 4x - 12 = 0$
$x^2 + 4x = 12$
Add $\left(\dfrac{4}{2}\right)^2 = 4$
$x^2 + 4x + 4 = 12 + 4$
$(x + 2)^2 = 16$
$x + 2 = 4; \quad x + 2 = -4$
$x = 2; \quad\quad x = -6$
$\{-6, 2\}$

14. $x^2 - x - 6 = 0$
$x^2 - x = 6$
Add $\left(\dfrac{-1}{2}\right)^2 = \dfrac{1}{4}$
$x^2 - x + \dfrac{1}{4} = 6 + \dfrac{1}{4}$
$\left(x - \dfrac{1}{2}\right)^2 = \dfrac{25}{4}$
$x - \dfrac{1}{2} = \dfrac{5}{2}; \quad x - \dfrac{1}{2} = \dfrac{-5}{2}$
$x = 3; \quad\quad x = -2$
$\{-2, 3\}$

15. $x^2 - 2x + 1 = 0$
$x^2 - 2x = -1$
Add $\left(\dfrac{-2}{2}\right)^2 = 1$
$x^2 - 2x + 1 = -1 + 1$
$(x - 1)^2 = 0$
$\{1\}$

16. $x^2 + 4x + 4 = 0$
$x^2 + 4x = -4$
Add $\left(\dfrac{4}{2}\right)^2 = 4$
$x^2 + 4x + 4 = -4 + 4$
$(x + 2)^2 = 0$
$\{-2\}$

17. $x^2 + 9x + 20 = 0$
$x^2 + 9x = -20$
$x^2 + 9x + \dfrac{81}{4} = -20 + \dfrac{81}{4}$
$\left(x + \dfrac{9}{2}\right)^2 = \dfrac{1}{4}$
$x + \dfrac{9}{2} = \dfrac{1}{2}; \quad x + \dfrac{9}{2} = \dfrac{-1}{2}$
$x = -4; \quad\quad x = -5$
$\{-4, -5\}$

18. $x^2 - x - 20 = 0$
$x^2 - x = 20$
$x^2 - x + \dfrac{1}{4} = 20 + \dfrac{1}{4}$
$\left(x - \dfrac{1}{2}\right)^2 = \dfrac{81}{4}$
$x - \dfrac{1}{2} = \dfrac{9}{2}; \quad x - \dfrac{1}{2} = \dfrac{-9}{2}$
$x = 5; \quad\quad x = -4$
$\{-4, 5\}$

19. $x^2 - 2x - 1 = 0$
$x^2 - 2x = 1$
$x^2 - 2x + 1 = 1 + 1$
$(x - 1)^2 = 2$
$x - 1 = \sqrt{2}; \quad x - 1 = -\sqrt{2}$
$x = 1 + \sqrt{2}; \quad x = 1 - \sqrt{2}$
$\{1 - \sqrt{2}, 1 + \sqrt{2}\}$

20. $x^2 + 3x - 1 = 0$
$x^2 + 3x = 1$
$x^2 + 3x + \dfrac{9}{4} = 1 + \dfrac{9}{4}$
$\left(x + \dfrac{3}{2}\right)^2 = \dfrac{13}{4}$
$x + \dfrac{3}{2} = \dfrac{\sqrt{13}}{2}; \quad x + \dfrac{3}{2} = \dfrac{-\sqrt{13}}{2}$
$x = -\dfrac{3}{2} + \dfrac{\sqrt{13}}{2}; \quad x = -\dfrac{3}{2} - \dfrac{\sqrt{13}}{2}$
$\left\{\dfrac{-3 + \sqrt{13}}{2}, \dfrac{-3 - \sqrt{13}}{2}\right\}$

21. $2x^2 = 4 - 3x$
$2x^2 + 3x = 4$
$x^2 + \dfrac{3}{2}x = 2$
$x^2 + \dfrac{3}{2}x + \dfrac{9}{16} = 2 + \dfrac{9}{16}$
$\left(x + \dfrac{3}{4}\right)^2 = \dfrac{41}{16}$
$x + \dfrac{3}{4} = \dfrac{\sqrt{41}}{4}; \quad x + \dfrac{3}{4} = \dfrac{-\sqrt{41}}{4}$
$x = -\dfrac{3}{4} + \dfrac{\sqrt{41}}{4}; \quad x = -\dfrac{3}{4} - \dfrac{\sqrt{41}}{4}$
$\left\{\dfrac{-3 + \sqrt{41}}{4}, \dfrac{-3 - \sqrt{41}}{4}\right\}$

22. $2x^2 = 6 - 5x$
$2x^2 + 5x = 6$
$x^2 + \dfrac{5}{2}x = 3$
$x^2 + \dfrac{5}{2}x + \dfrac{25}{16} = 3 + \dfrac{25}{16}$
$\left(x + \dfrac{5}{4}\right)^2 = \dfrac{73}{16}$
$x + \dfrac{5}{4} = \dfrac{\sqrt{73}}{4}; \quad x + \dfrac{5}{4} = \dfrac{-\sqrt{73}}{4}$
$x = -\dfrac{5}{4} + \dfrac{\sqrt{73}}{4}; \quad x = -\dfrac{5}{4} - \dfrac{\sqrt{73}}{4}$
$\left\{\dfrac{-5 + \sqrt{73}}{4}, \dfrac{-5 - \sqrt{73}}{4}\right\}$

23. $2x^2 + 4x = 3$
$x^2 + 2x = \dfrac{3}{2}$
$x^2 + 2x + 1 = \dfrac{3}{2} + 1$
$(x + 1)^2 = \dfrac{5}{2}$
$x + 1 = \sqrt{\dfrac{5}{2}}; \quad x + 1 = -\sqrt{\dfrac{5}{2}}$
$x = -1 + \sqrt{\dfrac{5}{2}}; \quad x = -1 - \sqrt{\dfrac{5}{2}}$
$\left\{-1 + \sqrt{\dfrac{5}{2}}, -1 - \sqrt{\dfrac{5}{2}}\right\}$

24. $3x^2 + x = 4$
$x^2 + \dfrac{1}{3}x = \dfrac{4}{3}$
$x^2 + \dfrac{1}{3}x + \dfrac{1}{36} = \dfrac{4}{3} + \dfrac{1}{36}$
$\left(x + \dfrac{1}{6}\right)^2 = \dfrac{49}{36}$
$x + \dfrac{1}{6} = \dfrac{7}{6}; \quad x + \dfrac{1}{6} = -\dfrac{7}{6}$
$x = 1; \quad x = -\dfrac{4}{3}$
$\left\{1, -\dfrac{4}{3}\right\}$

**Solution Key 2.4**

[2.4]

1. $x^2 - 3x + 4 = 0$
   $a = 1, \ b = -3, \ c = 4$

2. $5x^2 + x - 2 = 0$
   $a = 5, \ b = 1, \ c = -2$

3. $x^2 + 2x - 2 = 0$
   $a = 1, \ b = 2, \ c = -2$

4. $3x^2 - 3x + 1 = 0$
   $a = 3, \ b = -3, \ c = 1$

5. $2x^2 - x = 0$
   $a = 2, \ b = -1, \ c = 0$

6. $x^2 - 3x = 0$
   $a = 1, \ b = -3, \ c = 0$

7. $x^2 - 2x = 5$
   $x^2 - 2x - 5 = 0$
   $a = 1, \ b = -2, \ c = -5$

8. $3x^2 + 3x = 4$
   $3x^2 + 3x - 4 = 0$
   $a = 3, \ b = 3, \ c = -4$

9. $x^2 + 4x + 4 = 6$
   $x^2 + 4x - 2 = 0$
   $a = 1, \ b = 4, \ c = -2$

10. $x^2 - 2x + 1 = 7$
    $x^2 - 2x - 6 = 0$
    $a = 1, \ b = -2, \ c = -6$

11. $4x^2 + 4x + 1 = x^2 - x$
    $3x^2 + 5x + 1 = 0$
    $a = 3, \ b = 5, \ c = 1$

12. $3x^2 - 2 = x^2 + 6x + 9$
    $2x^2 - 6x - 11 = 0$
    $a = 2, \ b = -6, \ c = -11$

13. $x^2 - 5x + 4 = 0$
    $a = 1, \ b = -5, \ c = 4$
    $x = \dfrac{-(-5) \pm \sqrt{(-5)^2 - 4(1)(4)}}{2(1)}$
    $= \dfrac{5 \pm \sqrt{9}}{2}$
    $x = \dfrac{5 + 3}{2}; \ x = \dfrac{5 - 3}{2}$
    $\{4, 1\}$

14. $x^2 - 4x + 4 = 0$
    $a = 1, \ b = -4, \ c = 4$
    $x = \dfrac{-(-4) \pm \sqrt{(-4)^2 - 4(1)(4)}}{2(1)}$
    $= \dfrac{4 \pm 0}{2}$
    $x = \dfrac{4 + 0}{2}; \ x = \dfrac{4 - 0}{2}$
    $\{2\}$

15. $y^2 + 3y = 4$
    $y^2 + 3y - 4 = 0$
    $a = 1, \ b = 3, \ c = -4$
    $y = \dfrac{-3 \pm \sqrt{3^2 - 4(1)(-4)}}{2(1)}$
    $= \dfrac{-3 \pm \sqrt{25}}{2}$
    $y = \dfrac{-3 + 5}{2}; \ y = \dfrac{-3 - 5}{2}$
    $\{1, -4\}$

16. $y^2 - 5y = 6$
    $y^2 - 5y - 6 = 0$
    $a = 1, \ b = -5, \ c = -6$
    $y = \dfrac{-(-5) \pm \sqrt{(-5)^2 - 4(1)(-6)}}{2(1)}$
    $= \dfrac{5 \pm \sqrt{49}}{2}$
    $y = \dfrac{5 + 7}{2}; \ y = \dfrac{5 - 7}{2}$
    $\{6, -1\}$

**17.**
$$z^2 = 3z - 1$$
$$z^2 - 3z + 1 = 0$$
$$a = 1, \quad b = -3, \quad c = 1$$
$$z = \frac{-(-3) \pm \sqrt{(-3)^2 - 4(1)(1)}}{2(1)}$$
$$= \frac{3 \pm \sqrt{5}}{2}$$
$$z = \frac{3 + \sqrt{5}}{2}; \quad z = \frac{3 - \sqrt{5}}{2}$$
$$\left\{ \frac{3+\sqrt{5}}{2}, \frac{3-\sqrt{5}}{2} \right\}$$

**18.**
$$2z^2 = 7z - 2$$
$$2z^2 - 7z + 2 = 0$$
$$a = 2, \quad b = -7, \quad c = 2$$
$$z = \frac{-(-7) \pm \sqrt{(-7)^2 - 4(2)(2)}}{2(2)}$$
$$= \frac{7 \pm \sqrt{33}}{4}$$
$$z = \frac{7 + \sqrt{33}}{4}; \quad z = \frac{7 - \sqrt{33}}{4}$$
$$\left\{ \frac{7+\sqrt{33}}{4}, \frac{7-\sqrt{33}}{4} \right\}$$

**19.**
$$0 = 3x^2 - 5x + 1$$
$$a = 3, \quad b = -5, \quad c = 1$$
$$x = \frac{-(-5) \pm \sqrt{(-5)^2 - 4(3)(1)}}{2(3)}$$
$$= \frac{5 \pm \sqrt{13}}{6}$$
$$x = \frac{5 + \sqrt{13}}{6}; \quad x = \frac{5 - \sqrt{13}}{6}$$
$$\left\{ \frac{5+\sqrt{13}}{6}, \frac{5-\sqrt{13}}{6} \right\}$$

**20.**
$$0 = 2x^2 - x - 2$$
$$a = 2, \quad b = -1, \quad c = -2$$
$$x = \frac{-(-1) \pm \sqrt{(-1)^2 - 4(2)(-2)}}{2(2)}$$
$$= \frac{1 \pm \sqrt{17}}{4}$$
$$x = \frac{1 + \sqrt{17}}{4}; \quad x = \frac{1 - \sqrt{17}}{4}$$
$$\left\{ \frac{1+\sqrt{17}}{4}, \frac{1-\sqrt{17}}{4} \right\}$$

**21.**
$$5z + 6 = 6z^2$$
$$6z^2 - 5z - 6 = 0$$
$$a = 6, \quad b = -5, \quad c = -6$$
$$z = \frac{-(-5) \pm \sqrt{(-5)^2 - 4(6)(-6)}}{2(6)}$$
$$= \frac{5 \pm \sqrt{169}}{12}$$
$$z = \frac{5 + 13}{12}; \quad z = \frac{5 - 13}{12}$$
$$\left\{ \frac{3}{2}, \frac{-2}{3} \right\}$$

**22.**
$$13z + 5 = 6z^2$$
$$6z^2 - 13z - 5 = 0$$
$$a = 6, \quad b = -13, \quad c = -5$$
$$z = \frac{-(-13) \pm \sqrt{(-13)^2 - 4(6)(-5)}}{2(6)}$$
$$= \frac{13 \pm \sqrt{289}}{12}$$
$$z = \frac{13 + 17}{12}; \quad z = \frac{13 - 17}{12}$$
$$\left\{ \frac{5}{2}, -\frac{1}{3} \right\}$$

**23.**
$$x^2 + x = \frac{15}{4}$$
$$4x^2 + 4x - 15 = 0$$
$$a = 4, \quad b = 4, \quad c = -15$$
$$x = \frac{-4 \pm \sqrt{4^2 - 4(4)(-15)}}{2(4)}$$
$$= \frac{-4 \pm \sqrt{256}}{8}$$
$$x = \frac{-4 + 16}{8}; \quad x = \frac{-4 - 16}{8}$$
$$\left\{ \frac{3}{2}, \frac{-5}{2} \right\}$$

**24.**
$$\frac{y^2}{2} - \frac{y}{2} = 1$$
$$y^2 - y - 2 = 0$$
$$a = 1, \quad b = -1, \quad c = -2$$
$$y = \frac{-(-1) \pm \sqrt{(-1)^2 - 4(1)(-2)}}{2(1)}$$
$$= \frac{1 \pm \sqrt{9}}{2}$$
$$y = \frac{1 + 3}{2}; \quad y = \frac{1 - 3}{2}$$
$$\{2, -1\}$$

**Solution Key 2.5**

**25.**
$$\frac{x^2}{4} = 3 - \frac{x}{4}$$
$$x^2 + x - 12 = 0$$
$$a = 1, \quad b = 1, \quad c = -12$$
$$x = \frac{-1 \pm \sqrt{1^2 - 4(1)(-12)}}{2(1)}$$
$$= \frac{-1 \pm \sqrt{49}}{2}$$
$$x = \frac{-1 + 7}{2}; \quad x = \frac{-1 - 7}{2}$$
$$\{3, -4\}$$

**26.**
$$\frac{x^2 - 3}{2} + \frac{x}{4} = 1$$
$$4\left(\frac{x^2 - 3}{2}\right) + 4\left(\frac{x}{4}\right) = 4(1)$$
$$2x^2 + x - 10 = 0$$
$$a = 2, \quad b = 1, \quad c = -10$$
$$x = \frac{-1 \pm \sqrt{1^2 - 4(2)(-10)}}{2(2)}$$
$$= \frac{-1 \pm \sqrt{81}}{4}$$
$$x = \frac{-1 + 9}{4}; \quad x = \frac{-1 - 9}{4}$$
$$\left\{2, \frac{-5}{2}\right\}$$

## [2.5]

**1.** Let $n$ = one number
$\quad 13 - n$ = other number
$$n(13 - n) = 42$$
$$13n - n^2 = 42$$
$$n^2 - 13n + 42 = 0$$
$$(n - 7)(n - 6) = 0$$
$$n = 7; \quad n = 6$$
$$13 - n = 6; \quad 13 - n = 7$$
The numbers are 7 and 6.

**2.** Let $x$ = one number
$\quad x - 3$ = other number
$$x(x - 3) = 40$$
$$x^2 - 3x - 40 = 0$$
$$(x - 8)(x + 5) = 0$$
$$x = 8; \quad x = -5$$
$$x - 3 = 5; \quad x - 3 = -8$$
The numbers are 5 and 8 or $-5$ and $-8$.

**3.** Let $n$ = one odd integer
$\quad n + 2$ = other odd integer
$$n(n + 2) = 63$$
$$n^2 + 2n - 63 = 0$$
$$(n + 9)(n - 7) = 0$$
Since the numbers are positive,
$$n = 7$$
$$n + 2 = 9$$
The integers are 7 and 9.

**4.** Let $n$ = smaller integer
$\quad n + 2$ = larger integer
$$n(n + 2) = 48$$
$$n^2 + 2n - 48 = 0$$
$$(n + 8)(n - 6) = 0$$
Since the numbers are positive,
$$n = 6$$
$$n + 2 = 8$$
The integers are 6 and 8.

**5.** Let $y$ = the number
$$y + \frac{1}{y} = \frac{17}{4}$$
Multiply by $4y$:
$$4y^2 + 4 = 17y$$
$$4y^2 - 17y + 4 = 0$$
$$(4y - 1)(y - 4) = 0$$
$$y = \frac{1}{4}; \quad y = 4$$
The number is either 4 or $\frac{1}{4}$.

**6.** Let $n$ = the number
$$n - \frac{2}{n} = \frac{17}{3}$$
Multiply by $3n$:
$$3n^2 - 6 = 17n$$
$$3n^2 - 17n - 6 = 0$$
$$(3n + 1)(n - 6) = 0$$
$$n = -\frac{1}{3}; \quad n = 6$$
The number is either $-\frac{1}{3}$ or 6.

7.  $h = 64$, $h = 64t - 16t^2$
    Substitute 64 for $h$:
    $$64 = 64t - 16t^2$$
    $$16t^2 - 64t + 64 = 0$$
    $$16(t^2 - 4t + 4) = 0$$
    $$16(t - 2)^2 = 0$$
    $$t = 2$$
    It takes the ball 2 seconds to reach a height of 64 feet.

8.  $h = 0$, $h = 64t - 16t^2$
    Substitute 0 for $h$:
    $$0 = 64t - 16t^2$$
    $$16t^2 - 64t = 0$$
    $$16t(t - 4) = 0$$
    $$t = 0; \quad t = 4$$
    The ball is on the ground when it is thrown ($t = 0$), and it takes 4 seconds ($t = 4$) to return to the ground after it is thrown.

9.  $s = v_0 t + \frac{1}{2} g t^2$

    (a) Substitute 32 for $g$, 150 for $s$, and 20 for $v_0$:
    $$150 = 20t + 16t^2$$
    Multiply by $\frac{1}{2}$ and write in standard form:
    $$8t^2 + 10t - 75 = 0$$
    $$(2t - 5)(4t + 15) = 0$$
    $$t = \frac{5}{2}; \quad t = \frac{-15}{4} \quad \text{(time is not negative here)}$$
    It will take 2.5 seconds to fall 150 feet.

    (b) Starting from rest implies $v_0 = 0$.
    $$150 = 16t^2$$
    $$t^2 = \frac{150}{16}$$
    $$t = \pm\sqrt{\frac{150}{16}} = \pm\frac{5}{4}\sqrt{6} \quad \text{(time is not negative here)}$$
    It will take $\frac{5}{4}\sqrt{6}$ seconds to fall 150 feet, or approximately 3.06 seconds.

10. $s = v_0 t + \frac{1}{2} g t^2$. Substitute 280 for $s$, 24 for $v_0$, and 32 for $g$:
    $$280 = 24t + 16t^2$$
    Multiply by $\frac{1}{8}$ and write in standard form:
    $$2t^2 + 3t - 35 = 0$$
    $$(2t - 7)(t + 5) = 0$$
    $$t = \frac{7}{2}; \quad t = -5 \quad \text{(time is not negative here)}$$
    It will take 3.5 seconds.

11. Let $r = $ rate going; $\quad \frac{6}{r} = $ time going

    $r - 2 = $ rate returning; $\quad \frac{6}{r - 2} = $ time returning

    $$\frac{6}{r} + \frac{6}{r - 2} = \frac{5}{2}$$

Solution Key 2.5

Multiply by $2r(r-2)$:
$12(r-2) + 12r = 5r(r-2)$
$5r^2 - 34r + 24 = 0$
$(r-6)(5r-4) = 0$
$r = 6; \quad r = \frac{4}{5}$ (not possible if $r-2$ is rate returning)

The rate going is 6 mph; returning is 4 mph.

12. Let $x$ = speed of current

   $20 - x$ = speed of boat upstream; $\quad \frac{90}{20-x}$ = time upstream

   $20 + x$ = speed of boat downstream; $\quad \frac{75}{20+x}$ = time downstream

   $\frac{90}{20-x} = \frac{75}{20+x} + 3$

   Multiply by $(20-x)(20+x)$:
   $90(20+x) = 75(20-x) + 3(20+x)(20-x)$
   $x^2 + 55x - 300 = 0$
   $(x-5)(x+60) = 0$
   $x = 5; \quad x = -60 \quad (-60 \text{ is not practical})$

   The speed of the current is 5 mph.

13. Let $x$ = speed of current

   $18 - x$ = speed of boat upstream; $\quad \frac{30}{18-x}$ = time upstream

   $18 + x$ = speed of boat downstream; $\quad \frac{63}{18+x}$ = time downstream

   $\frac{30}{18-x} + 1 = \frac{63}{18+x}$

   Multiply by $(18-x)(18+x)$:
   $30(18+x) + (18-x)(18+x) = 63(18-x)$.
   $x^2 - 93x + 270 = 0$
   $(x-3)(x-90) = 0$
   $x = 3; \quad x = 90 \quad (90 \text{ is not practical})$

   The current speed is 3 mph.

14. Let $r$ = rate in traffic; $\quad \frac{10}{r}$ = time in traffic

   $r + 20$ = rate out of traffic; $\quad \frac{40}{r+20}$ = time out of traffic

   $\frac{10}{r} + \frac{40}{r+20} = \frac{3}{2}$

   Multiply by $2r(r+20)$:
   $20(r+20) + 80r = 3r(r+20)$
   $3r^2 - 40r - 400 = 0$
   $(r-20)(3r+20) = 0$
   $r = 20; \quad r = \frac{-20}{3}$ (no negative rates)

   The rates are 20 mph in traffic and 40 mph out of traffic.

**15.** Let $r$ = rate of morning train; $\dfrac{10}{r}$ = morning time

$r + 10$ = rate of evening train; $\dfrac{10}{r+10}$ = evening time

$\dfrac{10}{r} + \dfrac{10}{r+10} = \dfrac{5}{6}$ (Note: 50 minutes = $\dfrac{5}{6}$ hour)

$6(r+10)(10) + 6r(10) = r(r+10)(5)$

$5r^2 - 70r - 600 = 0$

Multiply by $\dfrac{1}{5}$:

$r^2 - 14r - 120 = 0$
$(r-20)(r+6) = 0$
$r = 20; \quad r = -6 \quad$ (no negative rates)

The rates are 20 mph in the morning and 30 mph in the evening.

## Unit 2 Review

**1.** $x^2 + 6 = 7x$
$x^2 - 7x + 6 = 0$
$(x-6)(x-1) = 0$
$x = 6; \quad x = 1$
$\{6,1\}$

**2.** $x^2 = 4x$
$x^2 - 4x = 0$
$x(x-4) = 0$
$x = 0; \quad x = 4$
$\{0,4\}$

**3.** $x(x+1) = 6$
$x^2 + x = 6$
$x^2 + x - 6 = 0$
$(x+3)(x-2) = 0$
$x = -3; \quad x = 2$
$\{-3,2\}$

**4.** $\dfrac{x^2}{4} + \dfrac{3x}{4} + \dfrac{1}{2} = 0$

$4\left(\dfrac{x^2}{4} + \dfrac{3x}{4} + \dfrac{1}{2}\right) = 4(0)$

$x^2 + 3x + 2 = 0$
$(x+2)(x+1) = 0$
$x = -2; \quad x = -1$
$\{-2,-1\}$

**5.** $5x^2 = 20$
$x^2 = 4$
$x = 2; \quad x = -2$
$\{2,-2\}$

**6.** $2x^2 - 10 = 0$
$2x^2 = 10$
$x^2 = 5$
$x = \sqrt{5}; \quad x = -\sqrt{5}$
$\{\sqrt{5}, -\sqrt{5}\}$

**7.** $\dfrac{3x^2}{4} - 12 = 0$

$\dfrac{3x^2}{4} = 12$

$3x^2 = 48$
$x^2 = 16$
$x = 4; \quad x = -4$
$\{4,-4\}$

**8.** $\dfrac{4x^2}{5} - 4 = 0$

$\dfrac{4x^2}{5} = 4$

$4x^2 = 20$
$x^2 = 5$
$x = \sqrt{5}; \quad x = -\sqrt{5}$
$\{\sqrt{5}, -\sqrt{5}\}$

**9.** $x^2 + 6x + 9 = (x+3)^2$

**10.** $x^2 - 8x + 16 = (x-4)^2$

**Solution Key   Unit 2 Review**

**11.** $x^2 + 3x + \dfrac{9}{4} = \left(x + \dfrac{3}{2}\right)^2$

**12.** $x^2 - x + \dfrac{1}{4} = \left(x - \dfrac{1}{2}\right)^2$

**13.** $x^2 - 6x - 7 = 0$
$x^2 - 6x = 7$
$x^2 - 6x + 9 = 7 + 9$
$(x - 3)^2 = 16$
$x - 3 = 4; \quad x - 3 = -4$
$x = 7; \quad\quad x = -1$
$\{7, -1\}$

**14.** $2x^2 + 5x - 3 = 0$
$x^2 + \dfrac{5}{2}x = \dfrac{3}{2}$
$x^2 + \dfrac{5}{2}x + \dfrac{25}{16} = \dfrac{3}{2} + \dfrac{25}{16}$
$\left(x + \dfrac{5}{4}\right)^2 = \dfrac{49}{16}$
$x + \dfrac{5}{4} = \dfrac{7}{4}; \quad x + \dfrac{5}{4} = -\dfrac{7}{4}$
$x = \dfrac{1}{2}; \quad\quad x = -3$
$\left\{\dfrac{1}{2}, -3\right\}$

**15.** $x^2 + 3x - 10 = 0$
$a = 1, \ b = 3, \ c = -10$
$x = \dfrac{-3 \pm \sqrt{3^2 - 4(1)(-10)}}{2(1)}$
$= \dfrac{-3 \pm \sqrt{9 + 40}}{2}$
$= \dfrac{-3 \pm 7}{2}$
$x = \dfrac{-3 + 7}{2}; \quad x = \dfrac{-3 - 7}{2}$
$x = 2 \quad\quad\quad x = -5$
$\{2, -5\}$

**16.** $y^2 - 2y = 8$
$y^2 - 2y - 8 = 0$
$a = 1, \ b = -2, \ c = -8$
$y = \dfrac{-(-2) \pm \sqrt{(-2)^2 - 4(1)(-8)}}{2(1)}$
$= \dfrac{2 \pm \sqrt{4 + 32}}{2}$
$= \dfrac{2 \pm 6}{2}$
$y = \dfrac{2 + 6}{2}; \quad y = \dfrac{2 - 6}{2}$
$y = 4 \quad\quad\quad y = -2$
$\{4, -2\}$

**17.** $\dfrac{y^2}{4} + y + \dfrac{3}{4} = 0$
$4\left(\dfrac{y^2}{4} + y + \dfrac{3}{4}\right) = 4(0)$
$y^2 + 4y + 3 = 0$
$a = 1, \ b = 4, \ c = 3$
$y = \dfrac{-4 \pm \sqrt{4^2 - 4(1)(3)}}{2(1)}$
$= \dfrac{-4 \pm \sqrt{16 - 12}}{2}$
$= \dfrac{-4 \pm 2}{2}$
$y = \dfrac{-4 + 2}{2}; \quad y = \dfrac{-4 - 2}{2}$
$\{-1, -3\}$

**18.** $x^2 + \dfrac{3}{2}x = \dfrac{1}{2}$
$2\left(x^2 + \dfrac{3}{2}x\right) = 2\left(\dfrac{1}{2}\right)$
$2x^2 + 3x = 1$
$2x^2 + 3x - 1 = 0$
$a = 2, \ b = 3, \ c = -1$
$x = \dfrac{-3 \pm \sqrt{3^2 - 4(2)(-1)}}{2(2)}$
$= \dfrac{-3 \pm \sqrt{9 + 8}}{4}$
$x = \dfrac{-3 + \sqrt{17}}{4}; \quad x = \dfrac{-3 - \sqrt{17}}{4}$
$\left\{\dfrac{-3 + \sqrt{17}}{4}, \dfrac{-3 - \sqrt{17}}{4}\right\}$

19. Let $x$ = the number

$\dfrac{1}{x}$ = its reciprocal

$$x + 2\left(\dfrac{1}{x}\right) = \dfrac{73}{6}$$
$$6x\left(x + \dfrac{2}{x}\right) = 6x\left(\dfrac{73}{6}\right)$$
$$6x^2 + 12 = 73x$$
$$6x^2 - 73x + 12 = 0$$
$$(6x - 1)(x - 12) = 0$$
$$6x - 1 = 0; \quad x - 12 = 0$$
$$x = \dfrac{1}{6}; \quad x = 12$$

The number is either $\dfrac{1}{6}$ or 12.

20. Let $r$ = original rate

$r + 20$ = increased rate

$$\text{Time} = \dfrac{\text{distance}}{\text{rate}}$$

$$\dfrac{\text{distance}}{\text{original rate}} + \dfrac{\text{distance}}{\text{increased rate}} = \text{total time}$$

$$\dfrac{20}{r} + \dfrac{30}{r + 20} = 1$$

$$r(r + 20)\left(\dfrac{20}{r} + \dfrac{30}{r + 20}\right) = r(r + 20)(1)$$

$$20(r + 20) + 30r = r(r + 20)$$
$$20r + 400 + 30r = r^2 + 20r$$
$$r^2 - 30r - 400 = 0$$
$$(r - 40)(r + 10) = 0$$
$$r = 40; \quad r = -10 \quad (-10 \text{ is impossible})$$

The man was traveling 40 mph originally.

# Appendix

1. $(x + 1)(x - 2) > 0$. The product is positive if the factors have the same sign.

   (a) If the factors have the same signs, then algebraically we have that

   $$(x + 1 > 0 \text{ and } x - 2 > 0) \text{ or } (x + 1 < 0 \text{ and } x - 2 < 0)$$

   These requirements are equivalent to

   $$(x > -1 \text{ and } x > 2) \text{ or } (x < -1 \text{ and } x < 2)$$

   However, if $(x > -1$ and $x > 2)$, then it is sufficient to write $x > 2$, since any real number greater than 2 is necessarily greater than $-1$. Similarly if $(x < -1$ and $x < 2)$, then it is sufficient to write $x < -1$. Hence the solution set is $\{x \mid x > 2 \text{ or } x < -1\}$.

   (b) The sign graph follows:

   ```
   (x - 2) ------- 0 + + +
   (x + 1) --- 0 + + + + + +
         |   |   |   |   |
        -5   0       5
   ```

   By inspection, the factors are both positive for all $x > 2$, and both negative for all $x < -1$. Hence the solution set is $\{x \mid x < -1 \text{ or } x > 2\}$.

2. $(x + 2)(x + 5) < 0$. The product will be negative if the factors have opposite signs.

   (a) If the factors have opposite signs, then algebraically we have that

   $$(x + 2 < 0 \text{ and } x + 5 > 0) \text{ or } (x + 2 > 0 \text{ and } x + 5 < 0)$$

   These requirements are equivalent to

   $$(x < -2 \text{ and } x > -5) \text{ or } (x > -2 \text{ add } x < -5)$$

   Now the requirement $(x < -2$ and $x > -5)$ can be written $(-5 < x < -2)$, but the requirement $(x > -2$ *and* $x < -5)$ can not be satisfied since there is no real number that is greater than $-2$ *and* less than $-5$. Hence the solution set is $\{x \mid -5 < x < -2\}$.

## Solution Key Appendix

(b) The sign graph follows:

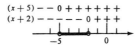

By inspection, the factors have opposite signs between $-5$ and $-2$, hence the solution set is $\{x \mid -5 < x < -2\}$.

3. $x(x-2) \leq 0$. The factors must have opposite signs, or at least one must be zero.
   (a) $(x \geq 0$ and $x-2 \leq 0)$ or $(x \leq 0$ and $x-2 \geq 0)$
   $(x \geq 0$ and $x \leq 2)$ or $(x \leq 0$ and $x \geq 2)$ (not possible)
   $\{x \mid 0 \leq x \leq 2\}$
   (b)
   $\{x \mid 0 \leq x \leq 2\}$

4. $x(x+3) \geq 0$. The factors must have the same sign, or at least one must be zero
   (a) $(x \geq 0$ and $x+3 \geq 0)$ or $(x \leq 0$ and $x+3 \leq 0)$
   $(x \geq 0$ and $x \geq -3)$ or $(x \leq 0$ and $x \leq -3)$
   $\{x \mid x \geq 0$ or $x \leq -3\}$
   (b)

   $\{x \mid x \geq 0$ or $x \leq -3\}$

5. $x^2 - 3x - 4 > 0$. Factor the left member.
   (a) $(x-4)(x+1) > 0$. Factors must have same sign.
   $(x-4 > 0$ and $x+1 > 0)$ or $(x-4 < 0$ and $x+1 < 0)$
   $(x > 4$ and $x > -1)$ or $(x < 4$ and $x < -1)$
   $\{x \mid x > 4$ or $x < -1\}$
   (b)

   $\{x \mid x > 4$ or $x < -1\}$

6. $x^2 - 5x - 6 \geq 0$. Factor the left member.
   (a) $(x-6)(x+1) \geq 0$. Factors must have same sign or be zero.
   $(x-6 \geq 0$ and $x+1 \geq 0)$ or $(x-6 \leq 0$ and $x+1 \leq 0)$
   $(x \geq 6$ and $x \geq -1)$ or $(x \leq 6$ and $x \leq -1)$
   $\{x \mid x \geq 6$ or $x \leq -1\}$
   (b)

   $\{x \mid x \geq 6$ or $x \leq -1\}$

7. $$\frac{2}{x} \leq 4$$
Multiply each member by $x^2$ $(x \neq 0)$:
$$(x^2)\frac{2}{x} \leq (x^2)4$$
$$2x \leq 4x^2$$
$$0 \leq 4x^2 - 2x$$
$$4x^2 - 2x \geq 0$$
$$2x(2x-1) \geq 0$$
$(x > 0$ and $2x - 1 \geq 0)$ or $(x < 0$ and $2x - 1 \leq 0)$
$\left(x > 0$ and $x \geq \frac{1}{2}\right)$ or $\left(x < 0$ and $x \leq \frac{1}{2}\right)$
$\left\{x \mid x \geq \frac{1}{2} \text{ or } x < 0\right\}$

8. $$\frac{3}{x-6} > 8$$
Multiply by $(x-6)^2$, $x \neq 6$:
$$(x-6)^2 \frac{3}{(x-6)} > 8(x-6)^2$$
$$3(x-6) > 8(x^2 - 12x + 36)$$
$$3x - 18 > 8x^2 - 96x + 288$$
$$0 > 8x^2 - 99x + 306$$
$$8x^2 - 99x + 306 < 0$$
$$(8x - 51)(x - 6) < 0$$
$(8x - 51 < 0$ and $x - 6 > 0)$ or $(8x - 51 > 0$ and $x - 6 < 0)$
$\left(x < \frac{51}{8}$ and $x > 6\right)$ or $\left(x > \frac{51}{8}$ and $x < 6\right)$ (not possible)
$\left\{x \mid 6 < x < \frac{51}{8}\right\}$

9. $$\frac{x}{x+2} > 4$$
$$(x+2)^2 \frac{x}{(x+2)} > 4(x+2)^2 \quad (x \neq -2)$$
$$x(x+2) > 4(x^2 + 4x + 4)$$
$$x^2 + 2x > 4x^2 + 16x + 16$$
$$0 > 3x^2 + 14x + 16$$
$$3x^2 + 14x + 16 < 0$$
$$(3x + 8)(x + 2) < 0$$
$(3x + 8 < 0$ and $x + 2 > 0)$ or $(3x + 8 > 0$ and $x + 2 < 0)$
$\left(x < -\frac{8}{3}$ and $x > -2\right)$ (not possible) or $\left(x > -\frac{8}{3}$ and $x < -2\right)$.
$\left\{x \mid -\frac{8}{3} < x < -2\right\}$

10. $$\frac{x+2}{x-2} \geq 6$$
$$(x-2)^2 \frac{(x+2)}{(x-2)} \geq 6(x-2)^2 \quad (x \neq 2)$$
$$(x-2)(x+2) \geq 6(x^2 - 4x + 4)$$
$$x^2 - 4 \geq 6x^2 - 24x + 24$$
$$0 \geq 5x^2 - 24x + 28$$
$$5x^2 - 24x + 28 \leq 0$$
$$(5x - 14)(x - 2) \leq 0$$
$(5x - 14 \leq 0 \quad \text{and} \quad x - 2 > 0) \quad \text{or} \quad (5x - 14 \geq 0 \quad \text{and} \quad x - 2 < 0)$
$\left( x \leq \frac{14}{5} \quad \text{and} \quad x > 2 \right) \quad \text{or} \quad \left( x \geq \frac{14}{5} \quad \text{and} \quad x < 2 \right)$ (not possible)
$\left\{ x \mid 2 < x \leq \frac{14}{5} \right\}$

# INDEX

Completing the square, 26
Conditional equation, 2

Equations
 conditional, 2
 equivalent, 2
 first-degree, 1
 member of, 2
 quadratic, 22
 root of, 1
 second-degree, 22
 solution of, 1, 5
 standard form, 22
Equivalent equations, 2
Equivalent inequalities, 11

Factoring, 22
First-degree equations, 1
First-degree inequalities, 1

Identities, 2
Inequalities, 10
 equivalent, 11
 first-degree, 1
 quadratic, 34
 solution of, 10, 34
 solution set of, 10
 word problems, 20

Perfect square trinomial, 26

Quadratic equations, 22
Quadratic formula, 28
Quadratic inequalities, 34

Roots of equations, 1

Second-degree equations, 22
 standard form, 22
Sign graph, 34
Solutions of equations, 1, 5
 completing the square, 26
 factoring, 23
 quadratic formula, 28
Solution set
 equations, 1
 inequalities, 10

Word problems, 13, 30
 age, 15
 coin, 16
 consecutive-integer, 14
 general, 13
 inequalities, 20
 interest, 17
 uniform-motion, 18

# SUMMARY  Volume 3: Equations and Inequalities in One Variable

The section where each item is first introduced is shown in parentheses. All variables denote real numbers.

## Symbols

**P(x), Q(x), R(x), etc.**   Names of polynomials (1.1)

## Properties

**1.** The equations $P(x) = Q(x)$ and each one of

$P(x) + R(x) = Q(x) + R(x)$
$P(x) - R(x) = Q(x) - R(x)$

$P(x) \cdot R(x) = Q(x) \cdot R(x) \quad (R(x) \neq 0)$
$\dfrac{P(x)}{R(x)} = \dfrac{Q(x)}{R(x)} \quad (R(x) \neq 0)$

are equivalent.

An equivalent equation is produced by

(a) adding the same polynomial to, or subtracting the same polynomial from, each member of an equation, or

(b) multiplying or dividing each member of an equation by the same polynomial representing a nonzero number. (1.1)

**2.** The inequalities $P(x) < Q(x)$ and each one of

$P(x) + R(x) < Q(x) + R(x)$
$P(x) - R(x) < Q(x) - R(x)$

$P(x) \cdot R(x) < Q(x) \cdot R(x) \quad (R(x) > 0)$
$\dfrac{P(x)}{R(x)} < \dfrac{Q(x)}{R(x)} \quad (R(x) > 0)$

$P(x) \cdot R(x) > Q(x) \cdot R(x) \quad (R(x) < 0)$
$\dfrac{P(x)}{R(x)} > \dfrac{Q(x)}{R(x)} \quad (R(x) < 0)$

are equivalent.

An equivalent inequality is produced by

(a) adding the same polynomial to, or subtracting the same polynomial from, each member of an inequality, or

(b) multiplying or dividing each member of an inequality by the same positive number, or

(c) multiplying or dividing each member of an inequality by the same negative number and reversing the order of the inequality. (1.4)

**3.** If $ab = 0$ then $a = 0$ or $b = 0$ (2.1)

**4.** If $c \geq 0$, then the solution set of $x^2 = c$ is $\{\sqrt{c}, -\sqrt{c}\}$ (2.2)

**5.** For $a \neq 0$, the solution set of $ax^2 + bx + c = 0$ is

$$\left\{ \dfrac{-b \pm \sqrt{b^2 - 4ac}}{2a} \right\} \quad (2.4)$$